DEVELOPMENTS IN RUBBER TECHNOLOGY—1
Improving Product Performance

THE DEVELOPMENTS SERIES

Developments in many fields of science and technology occur at such a pace that frequently there is a long delay before information about them becomes available and usually it is inconveniently scattered among several journals.

Developments Series books overcome these disadvantages by bringing together within one cover papers dealing with the latest trends and developments in a specific field of study and publishing them within *six months* of their being written.

Many subjects are covered by the series, including food science and technology, polymer science, civil and public health engineering, pressure vessels, composite materials, concrete, building science, petroleum technology, geology, etc.

Information on other titles in the series will gladly be sent on application to the publisher.

LIBRARY, RO WESTCOTT
REGULATIONS FOR BORROWERS

1. Books are issued on loan for a period of 1 month and must be returned to the library promptly.

2. Before books are taken from the Library receipts for them must be filled in, signed, and handed to a member of the Library Staff. Receipts for books received through the internal post must be signed and returned to the Library immediately.

3. Readers are responsible for books which they have borrowed, and are required to replace any such books which they lose or damage. In their own interest they are advised not to pass on to other readers books they have borrowed.

4. To enable the Library Staff to deal with urgent requests for books, borrowers who expect to be absent for more than a week are requested either to arrange for borrowed books to be made available to the PA or Clerk to the Section, or to return them to the Library for safekeeping during the period of absence.

DEVELOPMENTS IN RUBBER TECHNOLOGY—1

Improving Product Performance

Edited by

A. WHELAN and K. S. LEE

*National College of Rubber Technology,
Holloway, London, UK*

APPLIED SCIENCE PUBLISHERS LTD
LONDON

APPLIED SCIENCE PUBLISHERS LTD
RIPPLE ROAD, BARKING, ESSEX, ENGLAND

British Library Cataloguing in Publication Data

Developments in rubber technology.—(Developments series).
1: Improving product performance
1. Elastomers 2. Rubber
I. Whelan, Anthony II. Lee, K S III. Series
678   TS1925

ISBN 0-85334-862-6

WITH 89 TABLES AND 75 ILLUSTRATIONS

© APPLIED SCIENCE PUBLISHERS LTD 1979

All rights reserved. No part of this publication may be reproduced, stored in a retrieval system, or transmitted in any form or by any means, electronic, mechanical, photocopying, recording, or otherwise, without the prior written permission of the publishers, Applied Science Publishers Ltd., Ripple Road, Barking, Essex, England

Printed in Great Britain by Galliard (Printers) Ltd., Great Yarmouth

PREFACE

The service life and performance of a rubber product are dependent on the rubber chosen as a basis, together with the types and quantities of compounding ingredients which are incorporated. Careful selection of the vulcanising system, protective agent, etc. may enable a rubber to give a satisfactory performance in conditions which ordinarily would be considered beyond its range.

Careful attention to the design of the rubber product is another important factor which influences its behaviour and affects its length of service.

It must be emphasised that this book is not intended as a complete guide to improving product performance but is intended to review the present position in the major areas indicated. It should, therefore, be of interest to those concerned primarily with the manufacture and use of rubber products.

All the contributors were chosen because they have extensive experience in their chosen fields. We should like to express our thanks to them and their companies for their participation in the preparation of this book. Grateful thanks are also tendered to the Governors of the Polytechnic of North London for allowing us to act as editors and for providing the facilities without which this book could not have been produced.

A. WHELAN
K. S. LEE

CONTENTS

Preface v

List of Contributors ix

1. Developments with Natural Rubber 1
   D. J. ELLIOTT

2. Special-purpose Elastomers 45
   G. C. SWEET

3. Vulcanisation Systems 105
   E. R. RODGER

4. Carbon Blacks 151
   A. I. MEDALIA, R. R. JUENGEL and J. M. COLLINS

5. Silane-treated Mineral Fillers in Rubbers 183
   E. P. PLUEDDEMANN and B. THOMAS

6. Plasticisers 207
   G. MORRIS

7. Protective Agents 227
   B. T. ASHWORTH and P. HILL

8. Principles of Product Design 249
   E. SOUTHERN

Index 279

LIST OF CONTRIBUTORS

B. T. ASHWORTH
> *Vulnax International Limited, P.O. Box 116, Delaunays Road, Blackley, Manchester M60 1EP, UK*

J. M. COLLINS
> *Cabot Carbon Limited, Stanlow, Ellesmere Port, South Wirral L65 4HT, UK*

D. J. ELLIOTT
> *The Malaysian Rubber Producers' Research Association, Brickendonbury, Hertford SG13 8NL, UK*

P. HILL
> *Vulnax International Limited, P.O. Box 116, Delaunays Road, Blackley, Manchester M60 1EP, UK*

R. R. JUENGEL
> *Cabot Corporation, Research Centre, Concord Road, Billerica, Massachusetts, USA*

A. I. MEDALIA
> *Cabot Corporation, Research Centre, Concord Road, Billerica, Massachusetts, USA*

G. MORRIS
> *Curtagil Limited, Fishponds Road, Wokingham, Berks RG11 2QL, UK*

E. P. PLUEDDEMANN
  Dow Corning Corporation, Midland, Michigan 48640, USA

E. R. RODGER
  Monsanto Technical Center, Parc Scientifique, Rue Laid Burniat, B-1348 Louvain-La-Neuve, Belgium

E. SOUTHERN
  National College of Rubber Technology, Holloway, London N7 8DB, UK

G. C. SWEET
  Elastomers Research Laboratory, Du Pont (U.K.) Limited, Maylands Avenue, Hemel Hempstead, Herts HP2 7DP, UK

B. THOMAS
  Dow Corning Limited, Barry, Glamorgan CF6 7YL, UK

Chapter 1

DEVELOPMENTS WITH NATURAL RUBBER

D. J. ELLIOTT

The Malaysian Rubber Producers' Research Association, Brickendonbury, Hertford, UK

SUMMARY

Natural rubber has a unique combination of low hysteresis, and therefore high resilience, at low strains and high strength (due to crystallisation) at high strains, enabling it to be used in unfilled or lightly filled compounds. This ability to crystallise when stretched results from its uniform molecular structure. Strain-induced crystallisation is also responsible for NR's resistance to fatigue by a crack growth mechanism, an important property in dynamic applications of rubber.

Developments in the production of raw NR, in vulcanisation systems and in processing characteristics all provide improvements in the performance of products. Standard Malaysian rubber is guaranteed to high standards of quality, and several constant-viscosity and oil-extended grades, including a new general-purpose grade, are now available. These are designed for easy processing, with economic benefits, and better reproducibility of properties. A deproteinised grade has been developed for special applications.

Sulphur is still the most widely used crosslinking agent for NR, but new sulphur vulcanisation systems and some non-sulphur systems have advantages in specific areas. Thus soluble-EV systems and soluble activators are valuable where low stress-relaxation, high resilience and good modulus reproducibility are especially important. Vulcanisation with urethane reagents provides outstanding reversion resistance and high fatigue resistance in aged vulcanisates. Peroxide/coagent and peroxide/bismaleimide crosslinking systems have delayed action and good reversion resistance which widens the possibilities of the non-sulphur vulcanisation of NR.

Knowledge of the effects of different sulphur vulcanisation systems on fatigue properties can help in the choice of the most suitable compound for dynamic applications and an important factor in the choice of carbon black is its structure rating, if creep and dynamic loss factor are to be minimised.

Tyre-tread compounds based on oil-extended NR have a combination of properties well suited to their use in winter tyres. Thus OENR compounds have better grip on ice than OESBR compounds and they wear less under most driving conditions during winter months. NR oil-extended at the latex stage improves the processing characteristics of retreading compounds based on NR/SBR blends.

The natural route to the polymerisation of cis-polyisoprene makes NR a less-attractive contender for structural modification compared with synthetic rubbers with which in-line polymerisation variations are possible. Nevertheless there are several methods of modifying NR, some of which are economically attractive at the present time. These include

1. *the production of different types of thermoplastic NR by blending NR with polyolefins and by grafting azo-tipped polystyrene molecules on to an NR backbone;*
2. *epoxidation of NR to give high-damping rubbers;*
3. *the addition of low concentrations of reactive groups on to the NR molecule to provide sites for further utilisation, such as crosslinking and coupling with silica fillers.*

1. INTRODUCTION

Among the general-purpose rubbers now available natural rubber (NR) maintains its leading position as the most suitable for applications requiring a combination of high resilience, high strength and fatigue resistance.

The capacity to provide a strong, resilient material over a wide range of hardness is a direct result of the molecular structure of NR.

The very uniform, high-*cis*-polyisoprene molecules readily crystallise when the rubber is stretched beyond a certain point producing a superior form of reinforcement. Crystallisation limits movement between neighbouring molecular chains and results in a large increase in hysteresis and ultimate strength. Thus NR does not require a reinforcing filler for high tensile strength and therefore it can be compounded without or with only moderate amounts of carbon black and mineral fillers, giving highly resilient and durable products.

Strain-induced crystallisation is also responsible for NR's resistance to failure by a crack growth mechanism even when the strains involved are high. The growth of a crack, arising from a surface imperfection or small cut, is arrested because crystallisation occurs at the tip of the crack where stresses are high. Similarly under dynamic deformations the growth of cracks is very slow, particularly where the strain cycle has a finite minimum to ensure that crystallised rubber at the crack tips does not melt, and components can often be designed so that their dynamic deformation is superimposed on a static one.

For many applications rubber must perform consistently over a long period and NR's durability is illustrated by numerous examples, such as bridge bearings which have been in service for more than 20 years[1] and are showing no signs of deterioration.

Natural rubber, as traditionally compounded with conventional sulphur recipes, is adequate for many products but some applications need rather special properties, like extra-low creep or stress relaxation and very reproducible stiffness. For example, rubber components are increasingly being used for vehicle suspension systems in replacement of steel as the spring or bearing material. For trouble-free service in this type of application a low-creep rubber is required.

The versatility of NR is an advantage especially to the designer of engineering components. It can be used in compression, torsion or simple shear and, unlike steel, its modulus may be varied over a wide range by the incorporation of fillers, for example Young's modulus from about 1 MPa to 15 MPa (145 lbf/in^2 to 2175 lbf/in^2).

The performance of NR has been improved in recent years by a number of developments. Some of these are now quite well established; others, of a more tentative nature, could lead to a wider spectrum of uses for NR.

2. MODERN FORMS OF NR

The precise molecular structure of NR, 100% *cis*-1,4-polyisoprene, is an example of an enzyme-controlled chemical process in building high-molecular-weight compounds to perfection.

The high green strength and tack of NR are associated with its molecular structure and numerous attempts have been made to emulate these useful characteristics in synthetic general-purpose rubbers.[2-7]

As with most raw materials the quality of NR also depends on the manner in which it is harvested and processed. In recent years producers have been particularly concerned with providing rubber with consistently good properties.

Traditionally NR has been marketed in a variety of types and grades including smoked and air-dried sheets, pale crepes and remilled crepes, all made into 'bareback' bales weighing 100 kg or more and painted with a coating mixture to prevent adhesion between bales during transit. These types are graded by visual inspection.[8]

Technically classified grades have also been available in limited quantities for many years but the introduction of the Standard Malaysian Rubber (SMR) scheme in 1965 was a major step forward in providing NR to guaranteed specifications. At the same time modern drying methods enabled more-easily handleable 33-kg bales of polythene-wrapped block rubber to be produced via the Heveacrumb or other comminutive procedures.

The proportion of Malaysian rubber produced under the SMR scheme has been rising since its inception and is now approximately 35%. Other rubber-producing countries have adopted very similar specification schemes.

SMR specifications are mandatory and conformance to them is guaranteed. Properties are determined by ISO test methods, including analysis for dirt, ash, nitrogen, volatile matter, plasticity and plasticity-retention index, and colour limit for SMR-L. Values of these properties for different grades are given in Table 1[9] which shows that grades derived directly from latex by deliberate coagulation have the lowest dirt and ash contents. These revised specifications, adopted in 1979, have lowered limits for nitrogen and volatile matter for all grades. They represent a total upgrading of SMR. Limiting the volatile matter to a maximum of 0·8% is an assurance that the rubber is adequately dried and this should reduce, if not eliminate, the occasional bale that is found to contain wet patches, which has been a sporadic problem in the past.

The plasticity retention index (PRI) is a measure of raw-rubber oxidisability, high values indicating greatest resistance to oxidation. It has been shown that some important properties of vulcanised NR, such as resistance to heat ageing and dynamic heat build-up, are related to PRI[10] although probably to a lesser extent than to the presence of an antioxidant. PRI is a quality-assurance parameter since maltreatment during processing of raw rubber generally leads to low PRI values.

2.1. Viscosity-Stabilised Grades[11]

Natural rubber contains a small proportion of aldehyde groups attached to the polyisoprene molecules. Polymerisation or condensation of these

TABLE 1

STANDARD MALAYSIAN RUBBER SPECIFICATIONS AS FROM JANUARY 1979

Parameter[a]	Latex				Sheet material, SMR-5	Blend, viscosity-stabilised, SMR-GP	Field-grade material		
	Viscosity-stabilised		Not stabilised						
	SMR-CV	SMR-LV[b]	SMR-L	SMR-WF			SMR-10	SMR-20	SMR-50
Dirt retained on 44-μm aperture, max. % weight	0·03	0·03	0·03	0·03	0·05	0·10	0·10	0·20	0·50
Ash content, max. % weight	0·50	0·50	0·50	0·50	0·60	0·75	0·75	1·00	1·50
Nitrogen content, max. % weight	0·60	0·60	0·60	0·60	0·60	0·60	0·60	0·60	0·60
Volatile matter, max. % weight	0·80	0·80	0·80	0·80	0·80	0·80	0·80	0·80	0·80
Wallace rapid plasticity—minimum initial value, P_0	—	—	30	30	30	—	30	30	30
Plasticity retention index, PRI, min. %	60	60	60	60	60	50	50	40	30
Colour limit, Lovibond Scale, max.	—	—	6·0	—	—	—	—	—	—
Mooney viscosity, ML 1 + 4, 100°C	—[c]	—[d]	—	—	—	—[e]	—	—	—
Cure[f]	R	R	R	R	—	R	—	—	—
Colour coding marker[g]	Black	Black	Light green	Light green	Light green	Blue	Brown	Red	Yellow
Plastic wrap colour	Transparent	Transparent	Transparent	Transparent	Transparent	Transparent	Transparent	Transparent	Transparent
Plastic strip colour	Orange	Magenta	Transparent	Opaque white	Opaque white	Opaque white	Opaque white	Opaque white	Opaque white

[a] Testing for compliance shall follow ISO test methods (see ref. 2). Additional producer control parameter: acetone extract 6–8% by weight.
[b] Contains 4 phr light, non-staining mineral oil.
[c] Three sub-grades, viz. SMR-CV50, SMR-CV60 and SMR-CV70 with producer viscosity limits at 45–55, 55–65 and 65–75 units respectively.
[d] One grade designated SMR-LV50 with producer viscosity limits at 45–55 units.
[e] Producer viscosity limits are imposed at 58–72 units.
[f] Cure information is provided in the form of a rheograph (R).
[g] The colour of printing on the bale identification strip.

groups results in the slow formation of a small number of crosslinks, manifested by a gradual increase in the viscosity of the rubber. This is usually referred to as storage hardening. Treatment of rubber at the production stage with a monofunctional amine such as hydroxylamine, to block off the aldehyde groups, prevents the occurrence of storage hardening.[12-14] The viscosity and molecular weight of rubber prepared in this way are stabilised at the level it has when freshly made. Since different clones of Hevea brasiliensis produce rubber having different viscosities, preblending of latex is normally carried out to provide this constant viscosity form of NR within specified viscosity limits. SMR-CV is now marketed in three viscosity ranges and SMR-LV, which contains an added 4% of light mineral oil, in one viscosity range (Table 2).

TABLE 2

SUB-GRADES OF SMR-CV AND SMR-LV AND PRODUCER LIMITS ON MOONEY VISCOSITY MANDATORY FROM JANUARY 1979

Grade	Mooney viscosity, ML $1 + 4$, $100°C$
CV50	45–55
CV60	55–65
CV70	65–75
LV50	45–55

The chief attribute of SMR-CV and SMR-LV is that premastication can be reduced or even eliminated. This not only reduces processing costs but can also lead to improved uniformity of mix viscosity and thereby to better reproducibility of modulus after vulcanisation.

A new grade of NR recently introduced[9] is called general-purpose NR (SMR-GP). It is expected to become a large-volume grade for the manufacture of tyres, belting and general mechanicals since it meets a demand for a general-purpose grade having consistent processing characteristics. It is viscosity-stabilised and is produced from a blend of latex-grade rubber (60%) with field coagulum (40%). Specifications are similar to those of SMR-10 (Table 1).

2.2. Cure Testing

With more widespread use of rheometers in the rubber industry for routine control checks of curing behaviour, it became clear that there is a need to provide data in the form of rheographs on the curing characteristics of NR. This rheometer test conveys more comprehensive information than the older MOD test and is also intended to regulate the selection and blending of the raw material at the production stage. In this way improved uniformity within grades is obtained. Rheographs, obtained on a Monsanto Rheometer R100 model using ACS1 test mixes, are now supplied with the test certificates for SMR-L, SMR-WF, SMR-CV, SMR-LV and SMR-GP grades, for the purpose of providing information to the consumer, although they are not part of the guaranteed specification.

2.3. Deproteinised Natural Rubber

A deproteinised grade of natural rubber (DPNR) is produced from fresh field latex by treatment with a proteolytic enzyme.[15] Most of the proteins and their degradation products are removed by the special manufacturing methods used, and other naturally occurring hydrophilic substances, including traces of inorganic salts, are also removed.

The affinity for water of normal grades of NR is recognised to be low when compared with, for example, polychloroprene or polyurethane rubber.[16] DPNR has a still lower water affinity and is therefore a good grade for cable insulation. It has also been found useful for some applications requiring the lowest attainable creep properties, because absorption

TABLE 3

DPNR SPECIFICATION

Dirt, max. % weight	0·015
Ash, max. % weight	0·15
Nitrogen, max. % weight	0·15
Volatile matter, max. % weight	0·50
PRI, min. %	60
Mooney viscosity, ML 1 + 4, 100°C	50 ± 5

of water will itself cause creep, particularly of relatively thin sections of rubber. DPNR is a purified grade of NR as shown by the low limits for dirt, nitrogen, etc. (Table 3). It does contain, however, similar amounts of hydrophobic non-rubber substances as normal unpurified grades. Therefore, for critical applications in the medical field a post-vulcanisation extraction stage is usually required.

3. VULCANISATION SYSTEMS

The inherent elasticity of rubber is only realised when the linear molecules are crosslinked into a three-dimensional network by vulcanisation. The physical state changes from mainly plastic to highly elastic when crosslinks are inserted.

Sulphur vulcanisation is still the most widely used method of crosslinking NR. However, there have been developments in both sulphur and non-sulphur vulcanising systems which give substantial improvements in the properties and performance of NR components.[17]

3.1. Sulphur Systems

The amount of sulphur used in NR compounds, other than ebonite, can be varied from about 3·5 phr down to only 0·4 phr. Sulphur donors can be used to replace part or the whole of the elemental sulphur. As the amount of sulphur is reduced an optimum crosslink density is maintained by increasing the accelerator concentration, for this increases the efficiency with which sulphur reacts to form crosslinks.

A general division can be made into conventional systems which contain a much higher concentration of sulphur than accelerator, and efficient vulcanisation (EV) systems in which the concentration of sulphur is considerably lower than that of accelerator. Vulcanisation systems which are intermediate in composition between EV and conventional systems are often referred to as semi-EV. The precise dividing lines are, of course, arbitrary but the compositions given in Table 4 are a useful guide.

TABLE 4

SULPHUR (S) AND ACCELERATOR (A) CONCENTRATIONS AND A/S RATIOS OF DIFFERENT VULCANISATION SYSTEMS FOR NR

	$S,^a$ phr	A, phr	A/S
Conventional	2·0–3·5	1·2–0·4	0·1–0·6
Semi-EV	1·0–1·7	2·5–1·2	0·7–2·5
EV	0·4–0·8	5·0–2·0	2·5–12

^{a} Including the available sulphur from sulphur donors.

The ratio of any given accelerator to sulphur largely determines the types of sulphur crosslinks formed and also the amount of sulphur which combines with the rubber in non-crosslink structures such as cyclic sulphides.

Low ratios favour polysulphidic crosslinks with a relatively high level of cyclic sulphides while high ratios lead to crosslinks that are predominantly monosulphides, and cyclic sulphide formation is much reduced, though not eliminated. Time and temperature of vulcanisation also affect these features. Thus with all sulphur systems a higher curing temperature produces more cyclic sulphides and consequently fewer crosslinks, and therefore lower modulus. Also with conventional and semi-EV systems the number of sulphur atoms per crosslink diminishes with increasing cure time, from poly- to di- and perhaps a few mono-.[18] At the same time breakdown of polysulphidic crosslinks, with insertion of the liberated sulphur in cyclic sulphides, is the cause of reversion during vulcanisation.

Modification of the polyisoprene chains by cyclic sulphide groups, or indeed by any other group, can affect some properties. Thus the time-dependent hardening at low temperatures which results from crystallisation of the rubber is retarded, but excessive modification may increase susceptibility to oxidation.

Mono-sulphidic crosslinks are more stable to heat than polysulphidic crosslinks because the bond energy for rupture of the C–S bond is greater than that of the S–S bond.[19] This is one reason why EV systems give greater heat-ageing resistance and reversion resistance than conventional systems. Another reason, as explained above, is that main-chain modification, and consequently oxidisability, is greater with conventional systems even in the presence of antioxidants. The increased oxidisability has been traced to the presence of conjugated diene and triene groups which are formed concomitantly with cyclic sulphide groups.[20] A third reason for the improved ageing resistance given by EV systems is the relatively high concentrations of accelerator residues such as zinc dithiocarbamates or zinc mercaptobenzothiazole which function as antioxidants, probably by inactivating hydroperoxides, and behave synergistically with added antioxidants of the phenolic or amine types.

Typical vulcanisate structures at optimum cure times and the general properties discussed above are summarised in Table 5, and time-dependent changes in Young's modulus, resulting from crystallisation, for rubbers vulcanised with conventional and EV systems, stored unstrained at $-10°C$, are shown in Fig. 1.

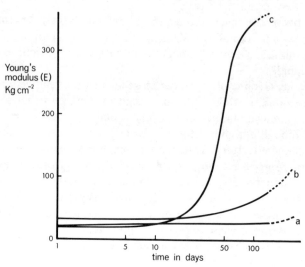

Fig. 1. Time-dependent changes in Young's modulus at $-10°C$ for some typical bridge-bearing vulcanisates. (a) NR conventional accelerated sulphur formulation. (b) crystallisation-resistant CR formulation. (c) NR soluble-EV formulation.

TABLE 5

VULCANISATE STRUCTURE AND PROPERTIES

	Conventional	Semi-EV	EV
Poly and disulphidic crosslinks, %	95	50	20
Monosulphidic crosslinks, %	5	50	80
Cyclic sulphide concentration	high	medium	low
Low-temperature-crystallisation resistance	high	medium	low
Heat-ageing resistance	low	medium	high
Reversion resistance	low	medium	high
Compression set, 22h at 70°C, %	30	20	10

From the foregoing it will be clear that EV systems have much to offer for applications requiring a good measure of heat resistance. An effect equivalent to overcuring may occur in a relatively thick component that is deformed dynamically in service and becomes overheated. The greater reversion resistance of an EV rubber can result in less overheating under these conditions. Also since EV systems are less sensitive to overcure it is possible to obtain more-uniform modulus in variable-thickness articles and especially throughout the thickness of bulky components that are cured for shorter times at higher temperatures.

Oxidative heat ageing of rubber is confined mainly to small components, thin sheets, and to near the surface of larger products in a hot environment, because its rate is controlled mainly by the diffusion rate of oxygen in rubber. At elevated temperatures oxygen combines with the surface layers more rapidly than it diffuses into the rubber and eventually a hard skin may be formed, particularly with conventionally vulcanised rubber but less readily with EV rubber.

A number of disadvantages of rubber cured with EV systems, especially those containing accelerators in concentrations well beyond their limiting solubilities in NR, have been identified. These are

1. fairly high primary (physical) creep rates;
2. variable strength at elevated temperatures;
3. lower resilience than conventional sulphur vulcanisates;
4. lower fatigue resistance when flexed through zero strain.

3.2. Soluble-EV Systems

Soluble-EV systems[21,22] are designed to overcome some of these deficiencies. They contain sulphur and accelerators in concentrations that ensure their complete solubility in the rubber at ambient temperature, so that crystals do not separate out during storage of the unvulcanised compound. Up to 0·8 phr of sulphur will remain dissolved in NR with no tendency to bloom over long periods. The accelerators N-oxydiethylenebenzothiazole-2-sulphenamide (OBS), tetrabutylthiuram disulphide (TBTD) and N,N'-diphenylguanidine (DPG) are soluble in rubber up to or beyond concentrations likely to be required, but 2-mercaptobenzothiazole (MBT), 2,2'-dibenzothiazyl disulphide (MBTS), N-cyclohexylbenzothiazole-2-sulphenamide (CBS), tetramethylthiuram disulphide (TMTD), and tetraethylthiuram disulphide (TETD) have low solubilities at toom temperature.[22] The absence of crystals of sulphur or accelerator eliminates the possibility of flaws (local areas excessively crosslinked) being formed during vulcanisation and this leads to an improvement in strength and fatigue resistance at elevated temperatures.

The second principle of soluble-EV systems concerns the fatty acid. Zinc stearate and the zinc soaps of most other straight-chain aliphatic acids are only slightly soluble in rubber and separate out after vulcani-

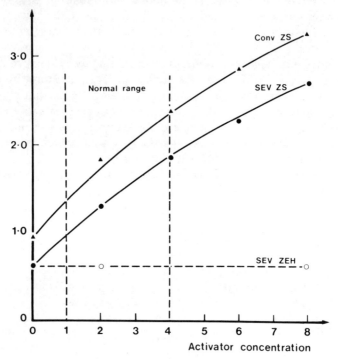

Fig. 2. Effects of activator concentration and curing system on stress-relaxation rate at 30°C of SMR-CV gum vulcanisates. Conv ZS = Conventional system (S, 2·5 phr; CBS, 0·6 phr) with zinc stearate. SEV ZS = Soluble-EV (S, 0·7 phr; OBS, 1·7 phr; TBTD, 0·7 phr) with zinc stearate. SEV ZEH = Soluble-EV with zinc 2-ethyl hexanoate. The modulus (MR100) of all vulcanisates was 0·73 MPa. Activator concentrations are expressed as the chemically equivalent stearic acid concentration in phr.

sation as crystals or micelles. These have been shown to be a major cause of primary creep and stress-relaxation in gum rubbers. The zinc soaps of branched-chain fatty acids such as zinc 2-ethyl hexanoate or zinc dimethyl hexanoate are much more soluble in rubber and when one of these is used in place of stearic acid a significant improvement in creep properties is obtained. Resilience is also improved.

The effects of curing system and zinc soaps on stress-relaxation rates are shown in Figs. 2 and 3 for gum and black-filled NR respectively.

Replacement of stearic acid by zinc 2-ethyl hexanoate can also be made in a conventional sulphur system with a consequent reduction in

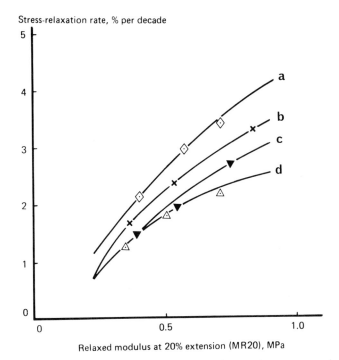

Fig. 3. Effects of activator, curing system and rubber grade on stress-relaxation rate of FEF(N550)-filled rubbers (20, 40 and 60 phr). (a) SMR-L conventional system (S, 2·5 phr; CBS, 0·6 phr) with stearic acid (2 phr). (b) SMR-L soluble-EV system (S, 0·7 phr; OBS, 1·7 phr; TBTD, 0·7 phr) with zinc 2-ethyl hexanoate (1 phr). (c) SMR-L conventional system with 2-ethyl hexanoate (1 phr). (d) DPNR, both conventional and soluble-EV systems with zinc 2-ethyl hexanoate (1 phr).

stress-relaxation rate, as shown in Fig. 3. Therefore, when there are good reasons for not using an EV system, for example if low-temperature-crystallisation resistance is an important requirement, a conventional system activated with a soluble zinc soap may be preferable.

3.3. Vulcanisation with Urethane Reagents[23]

A novel class of vulcanising agents which are basically diurethanes can be used to crosslink NR. The crosslinks are mainly of the urea type and their excellent stability provides good ageing and reversion resistance plus good initial vulcanisate properties.

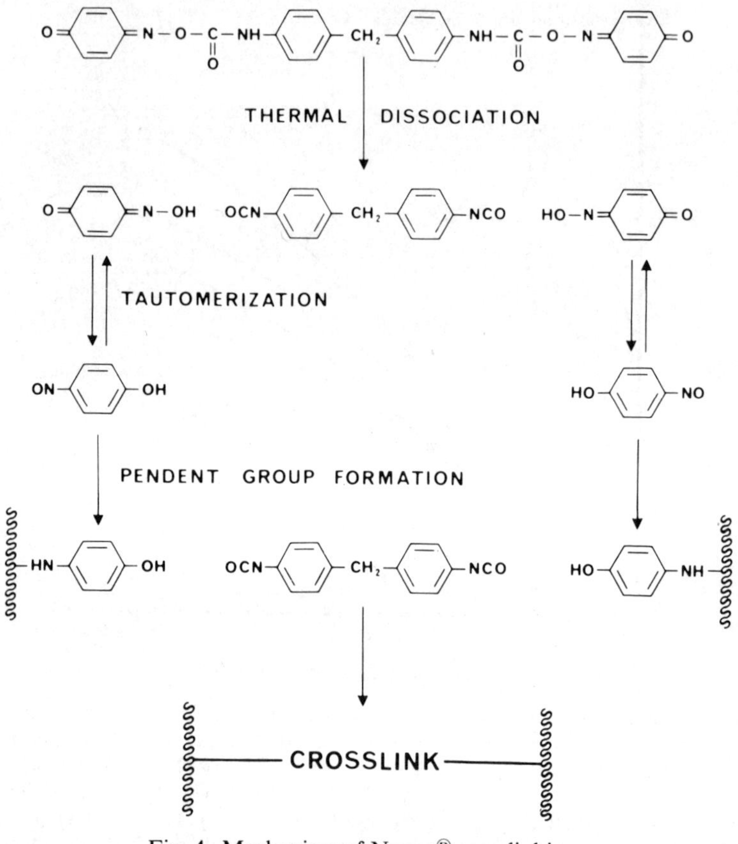

Fig. 4. Mechanism of Novor® crosslinking.

Urethane reagents are addition products of nitrosophenols and diisocyanates, stable at processing temperatures but which dissociate into their component species at vulcanising temperatures. The free nitrosophenol reacts with the rubber molecules to give pendent aminophenol groups as illustrated in Fig. 4. The aminophenol groups are then crosslinked by the liberated diisocyanate.

Urethane reagents are now marketed under the trade name Novor®. Several types are available but one of these, Novor® 924, is the most-

---

® Novor is a registered trade name of Durham Chemicals, Birtley, County Durham, UK.

widely used because it provides the greatest reversion resistance and also forms the basis of the best mixed urethane/sulphur systems as discussed later.

Two other ingredients are important to urethane crosslinking systems. The first of these is a drying agent based on calcium oxide to prevent hydrolysis of the diisocyanate by traces of moisture in the rubber and the second is zinc dimethyldithiocarbamate (ZDMC) which is a catalyst for the reaction between nitrosophenol and rubber. Without ZDMC the all-urethane vulcanisation is very slow.

The physical properties of natural rubber vulcanised with Novor® 924 and with various sulphur systems are listed for comparison in Table 6. This shows that the urethane systems give outstanding reversion resistance at 180°C, and resistance to heat ageing of vulcanisates cured to optimum equals that given by the soluble-EV system. Resilience of the urethane vulcanisates is slightly lower than that given by sulphur systems and consequently heat build-up is higher under comparable test conditions. Higher running temperatures can easily be accommodated by urethane vulcanisates, however, because of their superior reversion resistance.

Urethane systems are slow curing even with ZDMC present. It has been found that mixed urethane/sulphur systems are faster curing, more economical to use and retain in good measure the desirable properties of all-urethane systems. The synergism of the mixed systems, which produce higher crosslink density or modulus than either system individually, allows a reduction to be made in the concentration of the reagents. The much faster cure rate of early mixed systems was accompanied by very short scorch times caused by the ZDMC setting off the sulphur portion of the cure. This problem is alleviated by replacing the ZDMC with tetramethylthiuram monosulphide (TMTM) and some recommended urethane/sulphur formulations are shown in Table 7.

Urethane and mixed urethane/sulphur systems provide two valuable characteristics not given by straight sulphur systems. These are

1. a marked increase in fatigue resistance of the rubber on ageing, as opposed to a decrease with most sulphur systems; and
2. the ability to withstand vulcanisation temperatures up to 200°C without loss of modulus and tensile strength when compared with the magnitude of these properties obtained at lower vulcanisation temperatures.

TABLE 6

PHYSICAL PROPERTIES OF NATURAL RUBBER WITH VARIOUS VULCANISING SYSTEMS

	Urethane reagent with TMQ^c/ZMBI^b	Urethane reagent with HPPD^d	Conventional sulphur	Semi-EV	Soluble-EV
Rubber/black masterbatch^a	159	159	159	159	159
Stearic acid	1	1	3	2·5	
Zinc 2-ethyl hexanoate					1·5
ZMBI^b	2				
TMQ^c	2				
HPPD^d		2	2	2	2
Novor® 924	6·7	6·7			
ZDMC	2	2			
Caloxol	3	3			
CBS			0·5	0·8	
TMTD				0·4	
OBS^e					1·7
TBTD					0·7
Sulphur			2·5	1·2	0·7
Mooney scorch, t_5 at 120°C, min	24	13	25	13	23
Reversion after 1h at 180°C on rheometer, %	3	3	48	39	12

	Unaged properties				
Cure time/temperature, min/°C	40/160	40/160	40/140	30/140	40/150
Hardness, IRHD	69	72	65	65	60
MR100, MPa	1·9	1·8	2·0	2·2	1·6
M300, MPa	15	15	14	16	13
Tensile strength (TS), MPa	26	27	31	31	30
Elongation at break, %	455	460	545	495	560
Resilience, Dunlop tripsometer, %	62	65	66	70	68
Tension fatigue, 0–100% extension, kc	100	101	258	121	102
Goodrich flexometer, heat build-up after 30 min at 23°C, °C	46	38	32	36	30
Compression set, 1 day at 70°C, %	24	23	33	16	14
	Aged properties				
Retention of TS after 7 days at 100°C, %	69	47	21	36	69

[a] SMR-5, 100; HAF black (N330), 50; process oil, 4; zinc oxide, 5.
[b] Zinc-2-mercaptobenzimidazole.
[c] Polymerised 2,2,4-trimethyl-1,2-dihydroquinoline, Flectol® H.
[d] N-(1,3-dimethylbutyl)-N-phenyl-p-phenylenediamine, Santoflex® 13.
[e] N-oxydiethylenebenzothiazole-2-sulphenamide, Santocure® MOR.

® Flectol, Santoflex and Santocure are registered trade names of Monsanto Corp., St. Louis, Missouri, USA.

TABLE 7
RECOMMENDED URETHANE/SULPHUR FORMULATIONS

Urethane system/conventional sulphur system ratio	Concentration of ingredients, phr			
	Novor® 924	TMTM	Sulphur	TBBS[a]
90/10	4·8	1·4	0·2	0·04
80/20	4·2	1·3	0·4	0·08
70/30	3·8	1·2	0·6	0·12
60/40	3·2	1·1	0·8	0·16
50/50	2·7	1·0	1·0	0·20

[a] N-t-Butylbenzothiazole-2-sulphenamide.

Figure 5 shows how the fatigue life of an 80/20 urethane/sulphur-cured NR increases after ageing at 100°C while that of a conventional sulphur vulcanisate falls to very low values. This property is of particular value in a number of products such as transmission belting, bushes and flexible couplings operating in a hot environment.

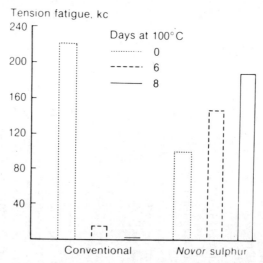

Fig. 5. Initial and aged tension fatigue lives of black-filled (SRF(N762), 5 phr) NR vulcanisates. Conventional (S, 2·5 phr; TBBS, 0·5 phr) with TMQ/HPPD antioxidants. Novor®/sulphur (Novor® 924, 4·2 phr; TMTM, 1·3 phr; S, 0·4 phr; CBS, 0·08 phr) with TMQ/ZMBI antioxidants.

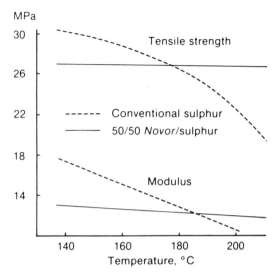

Fig. 6. Effect of vulcanisation temperature on tensile strength and modulus of black-filled (HAF(N330), 50 phr) NR vulcanisates. Conventional (S, 2·5 phr; TBBS, 0·5 phr). 50/50 Novor®/sulphur (Novor® 924, 3·2 phr; S, 1·2 phr; TMTM, 1·3 phr; TBBS, 0·24 phr).

Not all mixed urethane/sulphur systems have the feature described in (2) above to the full extent. The best in this respect is a 50/50 system as shown in Fig. 6 where it is compared with the conventional sulphur compound given in Table 6. The urethane/sulphur 50/50 system can be used to advantage in high-temperature vulcanisation, such as LCM curing at 190°C and injection moulding (180–190°C).

3.4. Delayed-action Peroxide Vulcanisation[24]

The use of organic peroxides as crosslinking agents for natural rubber was first investigated by Ostromislensky,[25] who used benzoyl peroxide, in 1915. However, it is only since dicumyl peroxide became available that this method of vulcanisation has been employed commercially. The vulcanisates have slightly inferior strength properties to those obtained with sulphur systems, but they have low creep and compression set and, with suitable antioxidants, excellent resistance to ageing. Due to the fact that the crosslinks are carbon–carbon bonds, the reversion resistance is also high.

The main technical limitations are a slow rate of cure and absence of delayed action. As with EV and urethane systems, curing may be carried out at high temperatures in order to reduce cure times, but then the short scorch time puts several restrictions on their utility. For example, the compound is too scorchy to be used satisfactorily in injection and often transfer moulding, and with compression moulding, backrinding is likely to occur in all but comparatively small components.

The mechanism of peroxide vulcanisation is very simple as shown by eqns. (1) to (3):

$$POOP \longrightarrow 2PO\cdot \tag{1}$$

$$PO\cdot + RH \longrightarrow POH + R\cdot \tag{2}$$

$$R\cdot + R\cdot \longrightarrow R-R \tag{3}$$

where POOP is an organic peroxide, RH represents polyisoprene and R—R represents crosslinked polyisoprene. Reaction steps (2) and (3) are much faster than (1) so that the rate-determining step is the thermal decomposition of the peroxide, a first-order reaction.

It is possible to impart some delay to the crosslinking reaction by introducing a radical scavenger such as N-nitrosodiphenylamine to remove $R\cdot$ or $PO\cdot$ radicals as they are formed. The drawback to this method is that removal of radicals lowers the crosslink yield and if more peroxide is used to compensate for this, most of the scorch delay is then lost. This is indicated by the rheometer traces shown in Figure 7.

If, however, a radical scavenger is used in conjuction with a peroxide coagent which increases the efficiency of crosslinking, an increase in peroxide concentration is not required and the delay period is main-

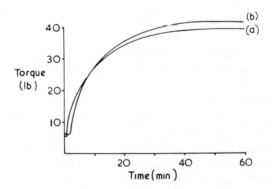

Fig. 7. Monsanto rheographs at 160°C. (a) SMR-5, 100; HAF(N330) black, 50; oil, 4; dicumyl peroxide, 2·5. (b) SMR-5, 100; HAF(N330) black, 50; oil, 4; dicumyl peroxide, 3·5; N-nitrosodiphenylamine, 0·4.

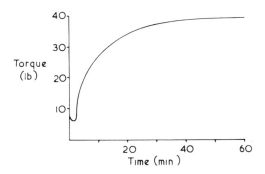

Fig. 8. Monsanto rheograph at 160°C. SMR-5, 100; HAF(N330) black, 50; oil, 4; dicumyl peroxide, 1·7; Saret® 500, 6.

tained. A commercially available coagent, Saret® 500,[26] is based on trimethylolpropanetrimethacrylate and a nitroso radical-capture reagent. When used in a natural rubber/dicumyl peroxide system it imparts some delay, as shown in Fig. 8.

It was found that the coagent more than compensated for the loss of crosslinking efficiency due to the radical scavenger, so a decrease in peroxide could be made and consequently the scorch time was increased even more.[24] This is shown in the plot of scorch time against Saret® 500 level in Fig. 9. The peroxide concentrations in phr required for constant modulus are given on the curve.

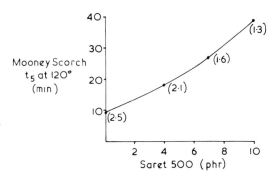

Fig. 9. Effect of Saret® 500 on Mooney scorch time for constant modulus peroxide/Saret® 500 compounds.

® Saret is a registered trade name of Sartomers Resins Inc., Essington, Pa, USA.

Radical scavengers can also be used to impact scorch delay to a peroxide-initiated bismaleimide crosslinking system. N,N'-m-phenylenebismaleimide will crosslink NR but requires initiation with a peroxide, MBTS or high-energy radiation.[27–29] The mechanism involves initiation, which is controlled by the peroxide (or MBTS) concentration, and crosslinking which is determined by the bismaleimide concentration. It follows that scorch delay can be adjusted by the addition of an appropriate amount of radical scavenger without significantly affecting the crosslink yield. Alternatively modulus can be varied by the maleimide concentration without affecting the scorch time. These predictions have been found to hold in practice.[24] The cure profiles of two compounds, one containing no retarder, the other containing 0·2 phr N-nitrosodiphenylamine, are shown in Fig. 10.

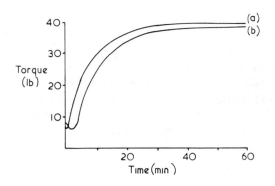

Fig. 10. Monsanto rheographs at 160°C. (a) SMR-5, 100; HAF(N330) black, 50; oil, 4; dicumyl peroxide, 0·6; HVA-2, 3·0. (b) as (a) with N-nitrosodiphenylamine, 0.2.

A further development of this system is the replacement of dicumyl peroxide with more-active peroxides that would normally be considered far too scorchy to be used in NR. The cure profile is then considerably improved over those normally associated with peroxide vulcanisation. That is, it exhibits a delay period followed by a rapid rate of cure. This is shown in Fig. 11 where the Trigonox® 29/40 bismaleimide/N-nitrosodiphenylamine curve (b) has a similar shape to that given by a

---

® Trigonox is a registered trade name of Akzo Chemie UK Ltd, London, UK.

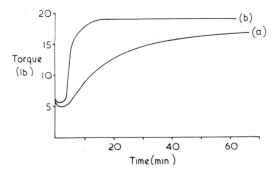

Fig. 11. Monsanto rheographs at 140°C. (a) SMR-5, 100; zinc oxide, 5; stearic acid, 2; S, 2·5; MBT, 0·6. (b) SMR-5, 100; Trigonox® 29/40, 1·2; HVA-2, 3·0; N-nitrosodiphenylamine, 0·3.

sulphur/sulphenamide system which for most purposes is better than the typical sulphur/MBT system (curve (a)) or the peroxide and peroxide/coagent systems (see Figs. 7 and 8).

The physical properties of NR crosslinked with the coagent and the bismaleimide types are very similar to those of the straight dicumyl peroxide system as shown in Tables 8 and 9 respectively.

TABLE 8

PROPERTIES OF NR VULCANISED WITH DICUMYL PEROXIDE AND DICUMYL PEROXIDE/SARET® 500

SMR-L	100	100
HAF-LS black (N326)	50	50
Process oil	4	4
Dicumyl peroxide[a]	2·5	1·3
Saret® 500		10
Mooney scorch, t_5 at 120°C, min	11	32
Cure time/temperature, min/°C	40/160	40/160
Tensile strength, MPa	21·5	23·5
MR100, MPa	2·5	2·6
Elongation at break, %	260	280
Resilience, Dunlop tripsometer, %	74	73
Compression set, 1 day at 70°C, %	3	6
Goodrich flexometer, heat build-up, °C	23	30

[a] Di-cup® R, Hercules (USA).

TABLE 9

PROPERTIES OF NR VULCANISED WITH DICUMYL PEROXIDE AND A DELAYED-ACTION BISMALEIMIDE SYSTEM

SMR-L	100	100
SRF black (N762)	25	25
Dicumyl peroxide[a]	2·5	0·6
N,N'-m-phenylene-bismaleimide[b]		3·0
N-nitrosodiphenylamine		0·2
Mooney scorch, t_5 at 120°C, min	10	52
Cure time/temperature, min/°C	40/160	30/160
Tensile strength, MPa	16	19
MR100, MPa	1·3	1·4
Elongation at break, %	340	355
Resilience, Dunlop tripsometer, %	87	85
Goodrich flexometer, heat build-up, °C	11	14

[a] Di-cup® R, Hercules (USA).
[b] HVA-2®, Du Pont (U.K.) Limited.

4. COMPOUNDING FOR ENGINEERING APPLICATIONS

The developments in vulcanisation systems which have been described are relevant to a number of products usually classified as engineering applications. These include engine and machinery mounts, transmission couplings, vehicle suspension springs and bushes, building mounts, bridge and rail-track bearings, and pipe seals for water and sewage.

High reproducibility of modulus is important in most of these products and here the flat curing characteristics of EV systems in conjunction with SMR-CV help to maintain consistency in a component being mass-produced.

Engine mounts are required to withstand higher temperatures than hitherto, which calls for the superior heat ageing and reversion resistance given by EV, peroxide or urethane systems. This will increase their serviceable temperature by about 20°C over that of conventionally vulcanised NR.

The property of low creep afforded by the soluble-EV system has proved beneficial in vehicle springs and shock-absorber bushes. For water seals, low stress relaxation under humid conditions is of paramount importance; therefore compounds based on DPNR vulcanised with a soluble-EV system would seem to be a sensible choice.

4.1. Fatigue Behaviour

The type of crosslink in vulcanised NR is one factor affecting the fatigue life of components. If the strain cycle goes through zero strain, conventional sulphur systems give longer fatigue lives than other vulcanisation systems, although this advantage is rapidly lost if the rubber is heat-aged as shown previously in Fig. 5. Rubber springs and mountings generally operate under a static load over which is superimposed a smaller dynamic load. In these non-relaxing, or finite minimum strain, conditions the fatigue lives of all NR vulcanisates are much greater than in relaxing fatigue. This is shown in Fig. 12[30] which also contrasts the

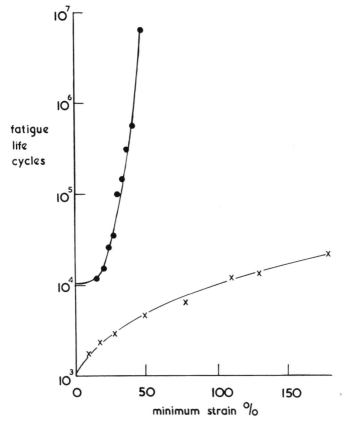

Fig. 12. Effect of minimum strain on fatigue life when maximum strain is 250%.
●, NR; ×, SBR.

behaviour of NR and SBR. Furthermore the inferior fatigue resistance under relaxing conditions given by EV systems compared with conventional systems disappears and may even be reversed when a finite minimum strain is introduced. Since cut growth behaviour is governed by stress crystallisation this behaviour is consistent with the fact that EV rubbers begin to crystallise at lower strains than conventional vulcanisates, perhaps because of their low content of cyclic sulphide groups.

Table 10 shows a comparison of the fatigue lives of several NR vulcanisates, each containing 45 phr FEF black and having comparable moduli, under both relaxing and non-relaxing tensile strains. For experimental reasons (fatigue lives can be very prolonged) the strains used are higher than those applicable to components in service. Nevertheless, this type of information should help in the choice of the most suitable compound for a particular product if the service strains involved can be ascertained.

TABLE 10

RING FATIGUE LIVES UNDER RELAXING AND NON-RELAXING TENSION (MEDIAN VALUES)

Vulcanisation system	kc to failure at the strain cycles	
	0–150%	50–200%
Conventional (S, 2·5 phr; CBS, 0·7 phr)	30	280
Semi-EV (S, 1·5 phr; CBS, 1·5 phr)	29	700
Soluble-EV (S, 0·8 phr; OBS, 2·0 phr; TBTD, 0·8 phr)	21	750

4.2. Effect of Carbon Black on Stress-relaxation Rate and Loss Factor

Two basic mechanisms govern the stress relaxation and creep of elastomers. One of these is caused by viscoelastic behaviour and results in a rate law in which stress relaxation is proportional to the logarithm of time. The other is caused by chemical changes in the polymer network and gives a rate law similar to that found in first-order chemical reactions. Chemical stress relaxation is dependent on the structure of the rubber and crosslinks, on presence of oxygen, etc., but is not markedly affected by the presence of carbon-black filler. On the other hand physical creep and stress-relaxation increase with amount of filler. Stress relaxation is related to the loss factor (tan δ) of the rubber which also increases with filler content, and low values of tan δ are preferable for most dynamic applications.[31]

The stress-relaxation rates of NR containing 60 phr of different particle size blacks are plotted in Fig. 13.[32] It is clear that the lowest rates are obtained with large-particle blacks. However the modulus of the rubber is also dependent on grade of carbon black and since modulus is a more important design parameter than the composition, it is appropriate to compare stress-relaxation rates at the same modulus, by varying the amounts of the different blacks.

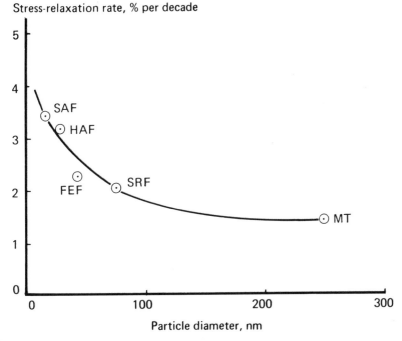

Fig. 13. Effect of filler particle size on stress-relaxation rate (DPNR/soluble-EV formulations, each with 60 phr black).

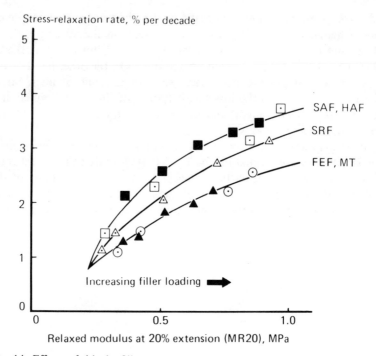

Fig. 14. Effect of black fillers on stress-relaxation rate (DPNR/soluble-EV formulation).

Figure 14 shows the variation of stress-relaxation rate with modulus for an NR/soluble-EV system containing five grades of carbon black. The points indicate different levels of each black to give a modulus (MR20) range from about 0·2 MPa to 1·0 MPa. The most significant feature is that although the very small particle size blacks, SAF and HAF, cause the highest relaxation rates, FEF black gives a lower relaxation rate than SRF, even though FEF has a smaller particle size than SRF. Relaxation rates at any one modulus are not, therefore, simply related to filler particle size. The reason for this is that the modulus of a filled rubber is not determined primarily by the particle size of the filler, but by its structure. FEF (N550) has a higher structure than the SRF (N762) and therefore less of it is required to achieve a given modulus. The effect on relaxation rate of this reduced loading more than compensates for the effect of decreased particle size.

The structure ratings of HAF and SAF blacks are not adequate to compensate for their very small particle sizes. Hence it is advantageous to choose a high-structure black in the medium particle size range in order to obtain low stress-relaxation rates. FEF (N550), GPF (N650) and SRF (N765) blacks fulfil these requirements.

As stated above rubbers having low stress-relaxation rates also have low dynamic loss factors (tan δ) and consequently low heat build-up properties, which is usually of importance in dynamic applications.

5. OIL-EXTENDED NATURAL RUBBER

The large-volume use of oil-extended SBR in passenger car tyres dates from the 1950s when it was discovered that oil-extended compounds had good wear properties and gave improved skid resistance. It was then shown that NR tread compounds could be extended with oil also, effecting cost savings and giving similar good skid resistance on wet roads. On ice, however, OENR compounds were found to be better than OESBR.[33]

Under certain conditions the wear resistance of OENR treads was greater than OESBR treads, but at other times, especially during summer months, it was worse. Relative wear ratings were shown to depend on a single parameter, the tyre surface temperature, irrespective of the conditions of use or the absolute rate of wear.[33,34]

Figure 15 shows the wear rating (the higher the rating the less the wear) of OENR relative to that of OESBR as a function of tyre surface temperature. These data were obtained from road trials with test vehicles under normal and severe driving conditions and with the MRPRA experimental trailer, which is particularly suited to the compilation of reliable data from relatively short test runs.[35] At tyre surface temperatures below 35°C OENR treads wear better than OESBR treads. During winter months in Great Britain tyre surface temperatures average between 20–25°C for short journeys and 30–35°C for long fast journeys on motorways.[34] In countries with colder climates, tyre surface temperatures will be lower and, consequently, the relative wear rating of OENR higher.

5.1. Winter Tyres

In northern and mountainous countries it has become popular to equip cars with special tyres for winter use to provide improved grip on snow and ice. They usually have a bold, open-tread pattern and are often

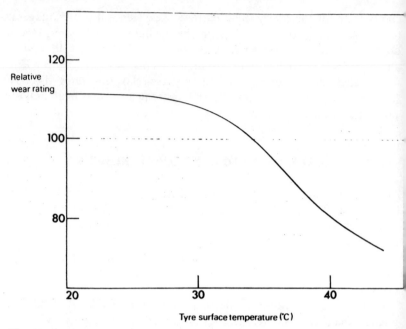

Fig. 15. Relative wear rating of OENR versus OESBR for tyre surface temperatures up to 45°C.

fitted with steel studs to increase the grip on ice. However, steel studs soon wear out and, because they have a disastrous effect on road surfaces, have been banned in some countries. Alongside tread pattern, therefore, the choice of rubber compound is important in the construction of winter tyres.[36,37]

If tyres were to be used exclusively on icy roads the highest skid resistance would be obtained with a natural-rubber tread, not oil-extended. This is indicated in Fig. 16[38] which shows that the coefficient of friction (the skid coefficient) of NR on ice is higher than that of SBR. Since winter tyres are also used on wet roads in less severe weather, their skid performance at temperatures above 0°C is also important, but here NR is inferior to SBR. However, OENR and OESBR have higher, but now equal, skid coefficients on wet surfaces as shown in Fig. 17[38] and although OENR is not as good as NR below 0°C it is still better than OESBR. Therefore for the optimum combination of good grip on wet and on icy roads the best tread rubber is OENR. As described in the

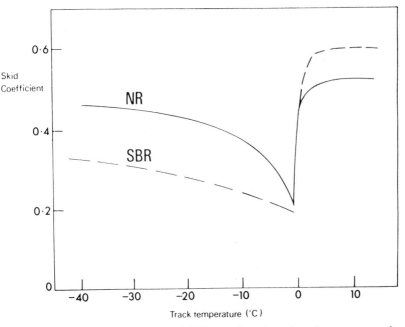

Fig. 16. Skid coefficients of NR and SBR as a function of track temperature for icy and, above 0°C, wet surfaces.

previous section such treads will have greater wear resistance than those based on OESBR.

The data of Figs. 16 and 17 were obtained with a Road Research Laboratory skid tester, on wet and ice-covered concrete slabs to simulate road surfaces, but road tests carried out in Sweden using the MRPRA trailer at temperatures from 0°C (melting ice) to −12°C have confirmed the relative skid ratings of OENR against OESBR.

5.2. OENR in Retreading Compounds

OESBR is the principal polymer used in the retreading of car and light-truck tyres, but in many processes, such as Orbitread® and extruder builders, a proportion of NR is usually needed to provide sufficient green adhesion of the tread stock to buffed carcase.

® Orbitread is a registered trade name of the American Machine and Foundry Company, Santa Ana, California, USA.

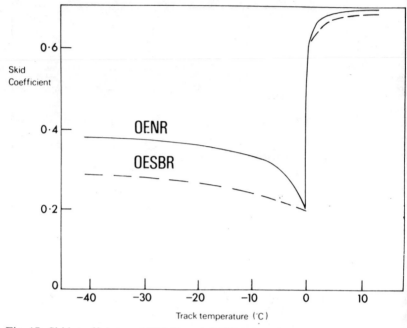

Fig. 17. Skid coefficients of OENR and OESBR as a function of track temperature for icy and, above 0°C, wet surfaces.

The NR must be premasticated, which is time- and energy-consuming, for otherwise the processability is adversely affected in that extrusion throughput is reduced.

It has now been demonstrated[39] that an NR/oil masterbatch (75% NR/25% aromatic oil), made by adding oil to latex before coagulation and requiring no premastication, can be used alone or blended with any desired proportion of OESBR, with or without additional oil, BR, NR or SBR, to produce compounds which process extremely well. Indeed it was found that as the proportion of OENR in OENR/OESBR/BR blends was increased, throughput using a ram extruder fitted with camelback die generally increased. Possible reasons for this behaviour are

1. viscosities decreased, and
2. mixing temperatures decreased,

as the proportion of OENR was increased. The lower mixing temperatures are believed to result in less carbon–rubber gel being formed and this gives improved flow properties.

6. MODIFICATION OF NATURAL RUBBER

The developments described so far are now fairly well established on a commercial basis. Some have helped NR to retain markets that might otherwise have gone to synthetic rubber, for example on account of inherently greater heat resistance. Others have enabled NR to make advances at the expense of synthetic rubber usage, a good example being winter tyres. In most cases, however, they have been the way to meeting stiffer specifications for new or existing products which are being made to operate under increasingly severe conditions. This section deals with more-recent topics, some of which could be commercially viable in the near future and others that are viewed as long-term developments.

The basic aims of this research are to devise new materials from and find new uses for NR.

6.1. Thermoplastic Natural-rubber Blends

Over the past few years there has been widespread interest in materials that can be moulded like thermoplastics but have a measure of the resilience, recovery or flexibility of vulcanised rubbers at the temperature of use.

One type of thermoplastic NR[40] falls into the broad class of materials based on physical blends of an elastomer and a thermoplastic resin. It is made by blending NR with crystalline polyolefins such as high-density polyethylene or isotactic polypropylene and has similar properties to blends of EPDM and polypropylene. They are generally hard materials at the interface between rubbers and rigid plastics.

Thermoplastic rubbers of the SBS block copolymer and segmented polyurethane types depend upon the presence of hard glassy or microcrystalline domains which function as crosslinks. With NR/polyolefin blends the crystalline polyolefin phase provides stiffness and reinforcement, and a small degree of crosslinking of the rubber phase prior to moulding further enhances stiffness and strength and, perhaps unexpectedly, improves moulding behaviour and surface appearance and reduces shrinkage (an effect perhaps similar to that given by superior processing NR, which is made from a blend of vulcanised and unvulcanised latex particles). The partial crosslinking of the NR phase is accomplished with an organic peroxide or other vulcanisation agent during preparation of the blend in an internal mixer. Thus a typical mixing procedure is to add polyolefin and NR to an internal mixer and masticate until the polyolefin melts and is blended with the rubber. The time required will depend upon the temperature of the mixer and the rotor speed and can be accomplished in 4 min in a BR Banbury heated with steam (50 psi pressure) running at speed 1 (77 rpm) and in 3 min in a pre-warmed Shaw K2A Intermix with full ram pressure and high speed (50 rpm). Dicumyl peroxide (typical amount 0·6 parts based on the NR content) is next added and mixing continued for a further 2·5 min. One part of antioxidant (based on total polymer) is then mixed in for 1 min and the batch dumped. The dump temperature should be 180–200°C. After sheeting on a two-roll mill the blend can be granulated to give an average particle size of 4–5 mm, suitable for hopper addition to extruder or injection moulding machine.

Injection moulding of thermoplastic NR/polyolefin blends has been studied[40] and it has been demonstrated that scrap can be re-used. Material passed ten times through the injection moulding machine showed only marginal deterioration of stiffness and tensile strength.

The stiffness of an NR/polyolefin blend depends on the relative proportions of the two polymers and to a lesser extent on the direction in which test pieces are cut relative to the direction of flow in the mould. Figure 18 shows the effect of polypropylene (PP) content on flexural modulus for blends from 75/25 NR/PP to pure PP. A practically useful range of thermoplastic NR blends is between 75/25 and about 15/85 NR/PP. The former is a material with similar properties to sole crepe (which relies on crystallised rubber for its reinforcement) whilst the latter is typical of high-impact-resistant grades of polypropylene.

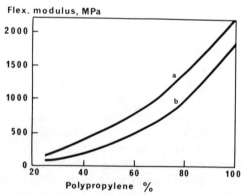

Fig. 18. Effect of polypropylene content on flexural modulus at 23°C of thermoplastic NR/polypropylene blends. (a) Radial direction; (b) tangential direction in square sheet mould filled from centre.

Between these limits are blends that are suitable for automotive applications such as flexible sight shields, rubbing strips and bumper components. These products form part of systems designed to prevent damage to the vehicle in minor accidents. The material must have sufficient stiffness for sight shield or bumper to support its own weight in all weather conditions from $-30°C$ to $+70°C$ and to withstand distortion during paint-baking processes at 120°C, where this is applicable.

The physical properties of NR/polyolefin blends given in Table 11 suggest they will meet these requirements of the car industry. They have a desirably small coefficient of stiffness with temperature when compared with competitive materials like polyurethanes and EPDM and they can be painted with flexible urethane paints after the normal methods of surface preparation. Incorporation of carbon black at the mixing stage renders them sufficiently antistatic to accept paint applied electrostatically. If they are not required to be painted they can be colour-pigmented but best weatherability is afforded by the inclusion of carbon black. They do not crack when exposed, stretched 20%, to high concentrations of ozone (100 pphm) or to full sunlight.

6.2. Chemical Modification of Natural Rubber

Several chemically modified forms of NR have been utilised in the past, but have now been largely replaced by synthetic materials. Examples are cyclised rubber, *cis–trans* isomerised rubber and chlorinated rubber.

Graft polymers of NR with polymethylmethacrylate are used in adhesives and for hard flexible mouldings, but although they have certain similarities to the NR/PP blends described above, they cannot be used as unvulcanised thermoplastic rubbers.

Recently attention has been concentrated on the addition of new chemical groups to the NR molecule to produce two main categories of material:

1. those with a relatively low molar concentration of active groups, so that elastic properties are unchanged but reactive sites are available for further utilisation;
2. those with high concentrations, so that rubbery properties are markedly changed.

An extension of (1) is where the added groups are of medium to high molecular weight, such as a linear polymer. A new type of graft polymer,

TABLE 11

PROPERTIES OF NR/PP AND NR/PP/PE BLENDS

Natural rubber, SMR-5L	65	60	50	40	65	60	50	40
Polypropylene, MFI 20 (190°C)	35	40	50	60	17.5	20	25	30
High-density polyethylene, MFI 5.5 (190°C)	—	—	—	—	17.5	20	25	30
Dicumyl peroxide[a]	0.39	0.36	0.30	0.24	0.39	0.36	0.30	0.24
Antioxidant[b]	1	1	1	1	1	1	1	1
Moulding properties								
Tensile strength, MPa								
a[c]	11.0	13.6	16.1	19.8	14.4	15.0	20.0	21.0
b[d]	10.0	10.6	14.8	16.7	12.6	13.0	14.8	14.0
Elongation at break, %								
a	65	55	60	115	55	55	40	60
b	405	330	475	540	400	465	495	495
Flexural modulus, MPa								
23°C a	400	500	620	880	300	440	620	840
23°C b	200	240	370	580	140	220	380	580
70°C a	150	160	240	300	90	150	200	290
70°C b	70	80	130	210	40	60	120	190
−30°C a	1200	1400	1900	2100	750	1100	1550	2100
−30°C b	600	750	1100	1650	400	600	850	1550
$\dfrac{\text{Flexural modulus at }-30°C}{\text{Flexural modulus at }+70°C}$	8.2	8.9	8.1	7.4	8.8	8.1	7.5	7.5

[a] e.g. Di-cup® R (Hercules, USA).
[b] 2,2-Methylene-bis(4-methyl-6-t-butylphenol) e.g. Antioxidant 2246® (Anchor Chemical Co. Ltd., Clayton, Manchester, UK).
[c] Radial direction.
[d] Tangential direction.

referred to as a comb graft, is then produced. The thermoplastic natural rubber described later has this type of structure.

6.3. Modifications via the 'ene' Reaction[41,42]

This reaction was chosen, in preference to free-radical and ionic reactions, because it was anticipated that addition to NR would occur efficiently avoiding undesirable side reactions such as cyclisation and double-bond isomerisation.

The 'ene' addition to the isoprenic double bond takes place as follows:

where $X=Y$ may be $-N=O$, $=C=O$, $=C=S$, or $-N=N-$.

Of these the azo 'ene' reaction, $-N=N-$, has proved to be one of the most efficient and therefore useful. (It may be noted that the mechanism of vulcanisation with urethane reagents, described earlier, involves an 'ene' addition of a $-N=O$ group to give a hydroxylamine, which undergoes further reaction to an amine.) The reactivity of the azo compound can be varied by changing the substituents on the nitrogen atoms, and a compound with the desired rate of addition to dry rubber (so that a homogeneous modification can be carried out by normal mixing procedures) was shown to be ethyl N-phenylcarbamoylazoformate, $EtO \cdot OC \cdot N : N \cdot CO \cdot NH \cdot Ph$ (ENPCAF). The reaction is complete in about 7 min at 110°C, a convenient time and temperature for a batch operation in an internal mixer.

A low concentration of ENPCAF groups has only a small effect on the elastic properties of NR but reduces the rate of crystallisation at $-26°C$, for example, by a factor of about ten for 1 mole % modification. Higher levels of modification have a pronounced effect on the glass transition temperature and the related temperature for minimum rebound, as shown in Table 12.

TABLE 12

EFFECT OF MODIFICATION WITH ENPCAF ON TEMPERATURE OF MINIMUM REBOUND

ENPCAF, mole %	Temperature of minimum rebound, °C
0	−32
2·59	−23
5·65	−11
9·35	0
16·25	66

Derivatives of ENPCAF having specific chemical functionality can also be attached to NR molecules, thereby providing reactive sites which may be utilised for crosslinking, interaction with fillers, adhesion to other materials, grafting, etc. Some examples of this approach are indicated below.

1. The difunctional ENPCAF reagent,

$$CH_2(C_6H_4NH \cdot CO \cdot N:N \cdot CO \cdot OEt)_2,$$

will crosslink NR directly. It is a very rapid reaction which might be utilised for prevulcanising latex.

2. ENPCAF derivatives containing alkoxysilane groups form a new type of rubber/filler coupling agent since they contain rubber-reactive and silica-reactive groups.[43] The triethoxysilylpropyl derivative has been named SILCAF. It has a number of advantages over mercaptosilane coupling agents. Thus it allows greater flexibility in the order of incorporation in the compound, it has less effect on scorch behaviour and it leads to greater improvements in physical properties, especially resilience, high strain modulus and abrasion resistance.

3. A hydroxyl-terminated polystyrene of controlled molecular weight is treated with phosgene followed by methyl carbazate to give a hydrazo terminal group. Oxidation of the latter yields an azo-terminated polystyrene:[44,45]

The azo polystyrene is reacted with NR, either in solution or by dry blending in a high shear mixer to give a comb graft copolymer:

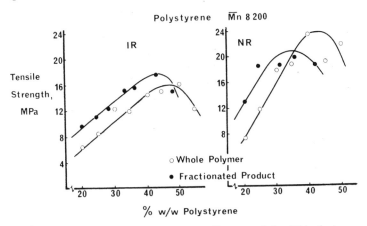

This material behaves as a thermoplastic rubber showing many of the characteristics of SBS block copolymers. Figure 19[46] shows the effect of polystyrene content on tensile strength for whole polymers and fractionated products of graft copolymers in which the backbone is NR and, for comparison, synthetic polyisoprene (IR). Grafting efficiency during dry mixing is known to be lower with NR, but in spite of this tensile strengths are generally higher than those obtained with the IR.

Fig. 19. Effect of polystyrene content on tensile strength for IR/polystyrene and NR/polystyrene graft copolymers.

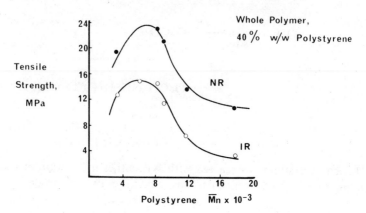

Fig. 20. Effect of molecular weight of polystyrene grafted chains on tensile strength for IR/polystyrene and NR/polystyrene graft copolymers.

There is an optimum molecular weight of the polystyrene for obtaining maximum tensile strength. This is shown in Fig. 20[46] for a copolymer containing 40% by weight of polystyrene. A pronounced peak occurs at a polystyrene molecular weight of 6000–8000, which is much lower than the molecular weight of the styrene blocks in SBS block copolymers, suggesting some differences in the morphology of block and graft copolymers.

6.4. Epoxidised Natural Rubber

Natural rubber, in latex form, reacts readily with peracids, the double bonds being converted into oxiranes:

The reaction must be carried out under controlled conditions, including keeping the temperature below 20°C and adding the peracid slowly, if secondary ring-opening reactions are to be avoided.

The reactivity of the epoxy groups towards difunctional agents such as diamines was found to be too low for them to be usefully employed as crosslinking sites,[47] although more-recent work has shown that their reactivity towards the amino acid glycine is much greater.[48,49] When epoxidation is carried to higher levels and vulcanisation is effected with sulphur or peroxide systems, the resulting modified rubber has some very interesting properties.[50] For example, there is an almost linear relationship between glass transition temperature (T_g) and degree of modification over the whole range from NR ($T_g = -72°C$) to 100 mole % epoxidised NR ($T_g = +5°C$).

Not surprisingly, damping also increases with level of epoxidation but stress-relaxation rate is hardly affected, remaining at the low value typical of NR. If, therefore, a peroxide or soluble-EV system is used for vulcanisation a soft rubber with high damping properties and low stress-relaxation rate will be obtained, a combination not possible with other high-damping rubbers.[51]

For applications where a hysteretic rubber is required to damp out vibrations it is desirable that the temperature bandwidth of damping should be wide enough to cover the service temperature range of the component. This can be achieved in the case of epoxidised NR by blending various levels of epoxidation, as shown in Fig. 21 where rebound resilience is plotted against temperature for NR, for materials having different levels of epoxidation and for a blend of three materials. It is believed that this behaviour results from a basic incompatibility of epoxidised rubbers having significantly different levels of epoxide groups, since otherwise a narrow bandwidth at an intermediate temperature range would be expected.

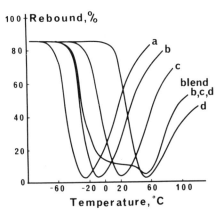

Fig. 21. Rebound resilience of NR and epoxidised NR vulcanisates. (a) NR; (b), (c) and (d) 25%, 50% and 90% epoxidised NR respectively.

Epoxidation of NR reduces its gas permeability and swelling in hydrocarbon oils as shown in Table 13 for gum vulcanisates.

TABLE 13

	Epoxidation, %			
	0	25	50	75
Relative permeability to air at 23°C	100	30	12	5
Swelling in ASTM No. 1 oil in 15 days, %	63·5	10·1	0·8	0·3
Swelling in ASTM No. 3 oil in 15 days, %	235	131	16	1·3

Highly epoxidised NR has some similarities to NBR but the tensile strengths of unfilled compounds are high (typically 29 MPa for a conventional sulphur/sulphenamide vulcanisate) compared with that of a soft NBR compound, probably because epoxidised NR undergoes stress-induced crystallisation.

ACKNOWLEDGEMENT

The author wishes to thank the board of MRPRA for permission to publish this paper.

REFERENCES

1. Anon. *Plastics Rubber Weekly* (778), 1979, 15.
2. GENDLER, T. R., BOGUSLAVSKII, D. B., LEVIT, G. M. and SAPRONOV, V. A. *Kauch. i. Rezina*, **31** (1), 1972, 11.
3. LUIJK, P. and RELLAGE, J. M. *Kautsch. Gummi Kunstst.*, **26** (10), 1973, 446.
4. BENDER, R. *Kautsch. Gummi Kunstst.*, **26** (12), 1973, 537.
5. SMIRNOV, V. P., KOVALEV, N. F., ESTRIN, A. S. and FURSENKO, A. V. *Kauch. i. Rezina*, **34** (9), 1975, 5.
6. BUCKLER, E. J., BRIGGS, C. J., HENDERSON, J. F. and LASIS, E. *Elastomerics*, **109** (12), 1977, 32.
7. BELGORODSKII, I. M., KOGAN, L. M., KROL, V. A., KOVALYOV, N. F. and SMIRNOV, V. P. *Rubber World*, **178** (1), 1978, 36.
8. *International Standards of Quality and Packing for Natural Rubber Grades* (*The Green Book*). The Rubber Manufacturer's Association, Inc. (USA), New York, 1969.
9. SMR Bulletin No. 9. Rubber Research Institute of Malaysia, Kuala Lumpur, 1978.

10. SMR Bulletin No. 4. Rubber Research Institute of Malaysia, Kuala Lumpur, 1966.
11. SMR Bulletin No. 8. Rubber Research Institute of Malaysia, Kuala Lumpur, 1970.
12. SEKHAR, B. C. *J. Polymer Sci.*, **48**, 1960, 133.
13. SEKHAR, B. C. British Patent 965,757, publ. 1964.
14. CHIN, P. S. *J. Rubber Res. Inst. Malaysia*, **22**, 1969, 56.
15. SMITH, J. F. British Patent 1,366,934, publ. 1974.
16. ROFF, W. F. and SCOTT, J. R. *Fibres, Films, Plastic and Rubber*, Butterworth, London, 1971.
17. ELLIOTT, D. J. and TIDD, B. K. In *Progress of Rubber Technology*, Vol. 34, PRI, London, 1974, 83.
18. CAMPBELL, D. S. *J. Appl. Polymer Sci.*, **14**, 1970, 1409.
19. BATEMAN, L. (Ed.). *The Chemistry and Physics of Rubber-like Substances*, Applied Science Publishers, London, 1963, 730.
20. PORTER, M. *The Chemistry of Sulphides*, A. V. Tobolsky (Ed.), Wiley, New York, 1968, 165.
21. ELLIOTT, D. J. *NR Technol.*, **3**, Part 3, 1972.
22. ELLIOTT, D. J. *Compounding Natural Rubber for Engineering Applications*, NR Technical Bulletin, MRPRA, Hertford, 1976.
23. BAKER, C. S. L. *Vulcanization with Urethane Reagents*, NR Technical Bulletin, MRPRA, Hertford, 1978.
24. CHOW, Y. L. and KNIGHT, G. T. *Proc. Intern. Rubber Conf.*, Brighton, 1977.
25. OSTROMISLENSKY, I. I. *J. Russ. Phys. Chem.*, **47**, 1915, 1467.
26. LOHR, J. E. and SAMPLES, R. *Rubber World*, **165**, October 1971, 47.
27. KOVACIK, P. and HEIN, R. W. *J. Am. Chem. Soc.*, **81**, 1959, 1187.
28. TAWNEY, P. O., WENISCH, W. J., VAN DER BERG, S. and RELYEA, D. I. *J. Appl. Polymer Sci.*, **8**, 1964, 2281.
29. MILLER, S. M., ROBERTS, R. and VALE, R. L. *J. Polymer Sci.*, **58**, 1962, 737.
30. LAKE, G. L. and LINDLEY, P. B. *Conference Proceedings*, Oxford, 1966, A. C. Strickland (Ed.), Institute of Physics and the Physical Society, London, 1966, 176.
31. GREGORY, M. J., METHERELL, C. and SMITH, J. F. *Proc. Intern. Rubber Conf.*, Brighton, 1977.
32. GREGORY, M. J. *NR Technol.*, **8**, Part 1, 1977, 1.
33. GROSCH, K. A. and SWIFT, P.McL., *Rubber Chem. Technol.*, **39**, 1966, 1656.
34. GROSCH, K. A. *J. Inst. Rubber Ind.*, **1**, 1967, 35.
35. SOUTHERN, E. *Tire Sci. Technol.*, **1** (1), 1973, 3.
36. SOUTHERN, E. *Proc. Intern. Rubber Conf.*, Kuala Lumpur, **IV**, 1975, 175.
37. SOUTHERN, E. and WALKER, R. W. *J. Inst. Rubber Ind.*, **6**, 1972, 249.
38. Anon. *OENR in Winter Tyre Treads*, NR Technical Bulletin, MRPRA, Hertford, 1968.
39. NEWELL, R., SMITH, I. F. and TAJUDDIN, B. K. *NR Technol.*, **9**, Part 3, 1978, 45.
40. CAMPBELL, D. S., ELLIOTT, D. J. and WHEELANS, M. A. *NR Technol.* **9**, Part 2, 1978, 21.
41. BARNARD, D., DAWES, K. and MENTE, P. G. *Proc. Intern. Rubber Conf.*, Kuala Lumpur, **IV**, 1975, 215.

42. PORTER, M. *Plastics Rubber: Materials Applications*, **3**, 1978, 32.
43. DAWES, K. and ROWLEY, R. J. *Plastics Rubber: Materials Applications*, **3**, 1978, 23.
44. CAMPBELL, D. S., LOEBER, D. E. and TINKER, A. J. *Polymer*, **19**, 1978, 1106.
45. CAMPBELL, D. S., LOEBER, D. E. and TINKER, A. J. *Polymer*, **20**, 1979, 393.
46. CAMPBELL, D. S. and TINKER, A. J., unpublished.
47. COLCLOUGH, T. *Trans. Inst. Rubber Ind.*, **38**, 1962, 11.
48. BURFIELD, D. R. and GAN, S. N. *J. Polymer Sci*, **13**, 1975, 2725.
49. BURFIELD, D. R., CHEW, L. C. and GAN, S. N. *Polymer*, **17**, 1976, 713.
50. GELLING, I. R. and SMITH, J. F., in press.
51. METHERELL, C. and SMITH, J. F. *Proc. Intern. Rubber Conf.*, Kuala Lumpur, **IV**, 1975, 140.

Chapter 2

SPECIAL-PURPOSE ELASTOMERS

G. C. Sweet

Du Pont (U.K.) Limited, Hemel Hempstead, Hertfordshire, UK

SUMMARY

This article reviews the chemistry, properties and applications of 15 special-purpose elastomers, including some developmental types which have not yet reached a commercial stage.

The terms of reference for selecting the elastomers reviewed are arbitrary and relate solely to end-use properties, and the utilisation of these properties to improve product performance.

Of the commercial products, chlorosulphonated polyethylene, silicones and fluorocarbons are dealt with in some depth. In addition, chlorinated polyethylene, epichlorhydrin, polyacrylic, ethylene acrylic, ethylene–vinyl acetate and polysulphide rubbers are reviewed.

In the area of semi-commercial products, there are sections dealing with phosphonitrilics, nitroso and carboxy nitroso, carborane and norbornene rubbers.

1. INTRODUCTION

Ever since the discovery of polymer chemistry earlier this century, the rubber chemist has strived to improve upon the properties of natural rubber in order to obtain elastomeric materials which are not affected by deleterious environments.

Heat, oils and solvents, oxidising chemicals and ozone are probably the most common elements which lead to the degradation of the natural-rubber molecule.

Hence synthetic elastomers have been developed which provide some, if not complete, protection against these factors. The introduction of neoprene in the early 1930s provided the industry with its first oil-resistant rubber.

The demands of the end-user have led to advances in polymer chemistry which now provide the industry with elastomeric materials which are resistant to continuous use at almost 300°C, are immune to attack by almost every known solvent and oxidising chemical and have saturated polymer backbones that resist ozone at exceptional concentration.

The improvement of one property, however, invariably leads to the deterioration of another. Highly oil-resistant polymers are usually very stiff and glass-like at low temperatures where another elastomer, with no resistance to oil, remains quite flexible. The skill of the rubber technologist and polymer chemist is stretched in order to provide end-use requirements that are usually a compromise. The universal elastomer is yet to be discovered.

In the meantime, we have a large number of elastomers that can be selected to provide special properties and the objective of this article is to highlight the advantages and disadvantages of each.

The special-purpose elastomer is usually more expensive than a general-purpose elastomer. This is partly the result of a supply/demand situation and partly due to the expensive feedstocks which have to be used to synthesise the product. Some of the elastomers currently regarded as special-purpose could well become general-purpose as demand increases and newer, less-expensive routes to synthesis are discovered. For example, ethylene acrylic rubber may well provide a new generation of relatively low-cost derivatives that might eventually replace polychloroprene as the world's most widely used rubber whose properties include oil resistance. Chlorosulphonated polyethylene and epichlorhydrin are already classed as general-purpose by ASTM designation.

On the other hand, it is unlikely that fluorocarbons, for instance, will ever be priced much lower than at present due primarily to feedstock price and availability.

In this review, ASTM designations are quoted by each heading where known. The reader is also referred to a recent paper on heat-resistant elastomers[1] which has been used as a basis for much of the data shown.

Finally, the omission of polypropylene oxide elastomer was intentional since this class of rubber, although limited in commercial use at present, would fit much better into a general purpose classification system.

2. CHLOROSULPHONATED POLYETHYLENE (CSM)

2.1. Manufacture
Chlorosulphonated polyethylenes are amorphous, vulcanisable elastic polymers marketed under the name Hypalon®. They are prepared by treating polyethylene in carbon tetrachloride solution with chlorine and sulphur dioxide gas yielding polymers containing 29–43% chlorine and 1–1·5% sulphur. Both low-density, branched chain and high-density, linear polyethylenes are utilised, CSM polymers from the former being used primarily for solution coating applications.[2]

2.2 Chemical Structure and Basic Properties
CSM may be represented structurally as

$$(-CH_2CH_2-)_x \quad (-CH_2CH-)_y \quad (-CH_2-CH-)_z$$
$$ | |$$
$$Cl SO_2$$
$$|$$
$$Cl$$

Values of x, y and z can be calculated from the chlorine and sulphur contents of the respective grades. The relation between molecular weight and intrinsic viscosity has been determined.[3] The degree of unsaturation of the polymers is negligible, hence they are highly resistant to heat, ozone and weathering. Being chlorinated, CSM vulcanisates are also resistant to oils and chemicals. Specific compounding techniques enhance these inherent properties. Owing to polarity of the CSM macromolecule, the dielectric properties of vulcanisates are only average. Special compounding techniques can provide vulcanisates with heat resistance to 150°C.[4]

2.3. Available Grades
Eight grades of Hypalon® are commercially available, as listed in Table 1.

2.4 Vulcanisation and Basic Compounding
High reactivity of the sulphonyl chloride crosslinking sites in CSM affords a wide choice of practical curing systems.[4,5] Current vulcanisation systems are summarised in Table 2. Recently, peroxide curing

---

® Hypalon is a registered trade name of Du Pont (U.K.) Limited.

TABLE 1

AVAILABLE GRADES OF HYPALON® (GRADES 20 AND 30 ARE LOW-DENSITY, BRANCHED CHAIN; ALL OTHER GRADES ARE HIGH-DENSITY, LINEAR TYPES.)

	Grade							
	20	30	40 soft	40	4085	45	48 soft	48
Chlorine content, %	29	43	35	35	35	25	43	43
Sulphur content, %	1·4	1·1	1·0	1·0	1·0	1·0	1·0	1·0
Specific gravity	1·14	1·26	1·18	1·18	1·18	1·07	1·27	1·27
Mooney viscosity at 100°C	30	30	45	55	85	40	62	77
Applications	Primarily solution coating		General purpose; extruded, calendered and moulded goods; coloured and black			Hard vulcanisates with moderate loading	Extruded, calendered and moulded goods with superior oil resistance	
						Unvulcanised compounds with good physical properties		

TABLE 2

VULCANISATION SYSTEMS FOR CSM

	Cure system						
	Metallic oxide + rubber accelerator			Metallic oxide + polyfunctional alcohol + rubber accelerator	Epoxy resin + rubber accelerator	Metallic oxide + bismaleimide	Metallic oxide + peroxide coagent
Magnesia	20	—	—	4	3	—	20
Litharge	—	20	25–40	—	—	30	—
Tetrone® A	2	0.75	1	2	1.5	—	—
MBTS	—	0.5	1	—	0.5	—	—
DOTG	—	—	—	—	0.25	—	—
Ni BD, NBC	—	3	3	—	—	1.5	—
HVA-2®	—	1	1	—	—	1.5	—
Pentaerythritol	—	—	—	3	—	—	—
Epoxy resin (Epikote 828®)	—	—	—	—	15	—	—
40% Dicumyl peroxide	—	—	—	—	—	—	8
Triallyl cyanurate	—	—	—	—	—	—	4
Hydrogenated wood resin	—	—	—	—	—	3	—
Type of crosslinks formed	Ionic and covalent			Ionic, ester and covalent	Epoxy and covalent	Complex	Complex
Outstanding property	Colour	Heat resistance	Water resistance	General purpose	Water resistance	Colour, Good electricals	Compression set

® HVA-2 (N,N' phenylene dimaleimide) is a registered trade name of Du Pont (U.K.) Limited.
Epikote 828 is a registered trade name of Shell Chemicals.

systems have been introduced.[6] Crosslinking with a metallic oxide/rubber accelerator curing system in the presence of acrylic type coagents has also been described.[7]

Crosslinking mechanisms of CSM with the various candidate vulcanisation systems are complex. The litharge/magnesia/Tetrone® A/Ni BD curing system appears to give the best overall heat resistance up to 150°C, as compared with other metallic oxide/rubber accelerator systems also stabilised with Ni BD. To ensure a minimum of covalent polysulphide crosslinks, the Tetrone® A content must be kept to a minimum consistent with adequate physical properties. Ni BD retards the decomposition of polysulphide crosslinks at elevated temperatures, which would result in the formation of additional sulphur crosslinks.

CSM can be compounded and processed on normal rubber equipment. Conventional fillers, carbon blacks, softeners, plasticisers and release agents are used. For optimum heat resistance, low-volatility plasticisers are necessary in association with fillers such as whiting, fine particle talcs and carbon blacks (other than MT, EPC and SAF) in addition to the correct crosslinking system.[8]

Vulcanisation is accomplished by the normal methods used in the rubber industry. In addition, CSM lends itself to several other methods of curing such as moisture, ammonia, radiation and UHF, and has been processed by LCM and fluidised bed techniques albeit with some difficulty.[9]

2.5 Applications

An American end-use analysis for 1977 indicated that 30% of CSM was utilised in coated fabrics and film sheeting, 40% in wire and cable, 4% in hose, 3% in tyres and the remainder in miscellaneous items, including rollers. Coatings based upon CSM, either on metals or fabrics, are used for chemical resistance. An example is acid-resistant tank linings. In the cable industry, CSM is used for its heat and oil resistance, notably as sheathing for nuclear power cables; off-shore oil rig and ships' cables and immersion heater flexibles; and also as insulation in coil-end flexibles, diesel–electric locomotive and welding cables. An associated application is as a base for semi-conductive compounds. CSM also forms a base for flexible magnetic strips.[10]

® Tetrone A (dipentamethylene thiuram tetrasulphide) is a registered trade name of Du Pont (U.K.) Limited.

In the construction industry, roofing and pit-liner systems, based upon unvulcanised, high green-strength CSM compounds, have been developed.[11-13] Prime requirements are excellent resistance to weathering and, for industrial effluent pit liners, good chemical resistance. A life of 20 years is predicted for CSM-based roofing systems.[12]

Chemical hoses lined with CSM are commercially produced. Automotive applications include ignition-cable sheathing, spark-plug boots and binders for cork-based gasketing.

The projected capacity increase for CSM suggests this polymer has substantial potential for further growth.

3. CHLORINATED POLYETHYLENE (CM)

3.1. Preparation and Grades

Chlorinated polyethylenes have been available since 1967[14] and have recently attracted increasing interest as general-purpose heat-, oil- and ozone-resistant elastomers for certain specific applications.

Various methods of manufacture for chlorinated polyethylenes have been reported.[15-18] A series of elastomers are commercially available based on the free-radical, random chlorination of high-density polyethylene in an aqueous slurry.[17] A range of five CM polymers is produced through variations in chlorine level, polyethylene molecular weight and distribution and reactor conditions.[19] These elastomers contain 36–42% chlorine and vary in molecular weight and inherent crystallinity to provide a balance in processing characteristics and physical properties. The Mooney viscosity range at 100°C is 50–90. They are supplied as free-flowing powders for processing on conventional rubber industry equipment.

A new elastomer produced by chlorination of low-density polyethylene has been announced.[20] The elastomer is offered in either granule or powder form for processing in a manner analogous to thermoplastics. Grades suitable for crosslinking to improve heat resistance and mechanical properties contain approximately 40% chlorine and are claimed to have good flame resistance.

Several papers have appeared covering the structural composition and properties of chlorinated polyethylenes.[21-27]

3.2. Vulcanisation

General techniques for processing and formulating commercial grades of CM have recently been reviewed.[28] Organic peroxides, often with a coagent to improve the rate and state of cure, are the preferred cross-linking agents. However, sulphur vulcanisation systems have been proposed for polymer previously heated with zinc oxide to introduce some C=C unsaturation in the polymer chain. Compared with peroxide cures, the process is more difficult to control and sulphur vulcanisates show less-satisfactory heat ageing and compression-set resistance.

As with peroxide cures generally, other compounding ingredients for CM should be selected with care since acidic and unsaturated materials can seriously impair crosslinking efficiency. Zinc derivatives should be avoided as these will initiate dehydrochlorination leading to polymer decomposition.

3.3. Compounding

Chlorinated polyethylene compounds should include a suitable heat stabiliser. Magnesium oxide, litharge, calcium oxide and monobasic silicate of white lead have been found particularly effective. Polymerised trimethyl dihydroquinoline antioxidants improve the long-term heat resistance of CM at temperatures to 150°C.

A number of studies on the effect of polymer chlorine level, compounding ingredients and compound contaminants on the heat resistance of CM have been reported.[29-31] CM vulcanisates are reversion-free at temperatures up to 150°C.[30]

3.4 Properties

The properties of CM vulcanisates have been compared with those of familiar oil-resistant elastomers, namely CR, NBR, ECO and CSM.[28,32] In general, chlorinated polyethylenes are characterised by good resistance to heat, oil, flame, chemicals, ozone and weathering. Commercial elastomers shown in Table 3 are claimed to perform at temperatures up to 162°C. Generally, as the polymer chlorine content increases, the oil, fuel and flame resistances are improved but heat resistance is lowered. With proper compounding, oil and fuel resistances of CM can approach the swelling values of typical NBR stocks. Studies have been reported on the rate of permeation of Freon® 12 through chlorinated polyethylene compounds, including blends with NBR and chlorobutyl.[33]

® Freon is a registered trade name of Du Pont (U.K.) Limited.

TABLE 3

DOW/BAYER CPE TYPES

Type	Chlorine content, %	Density, g cm^{-3}	Mooney viscosity, ML—1 + 4/120°C	Crystallinity
CM 3630	36	1·16	ca. 80	Very low
CM 3632	36	1·16	ca. 90	Low
CM 3610	36	1·16	ca. 30	Very low
CM 4230	42	1·25	ca. 90	Very low
CM 2552	25	1·10	ca. 150	Medium

3.5. Applications

Chlorinated polyethylenes find major applications in hose, wire and cable. Linings and covers for chemical transfer hoses have been produced from CM elastomers. These hoses are produced in sizes up to 10-cm internal diameter and are reinforced with rayon and galvanised steel wire. Resistance of CM vulcanisates to Freon® permeation makes this elastomer an attractive candidate for use in such applications as automotive air-conditioner hoses.

The electrical properties of CM in general limit its use as an insulant to low-voltage applications. However, the combination of high-temperature stability and good resistance to oils, chemicals, ozone and flame makes CM of particular interest for certain cable constructions. Patents claiming blends of CM with other elastomers for insulating[34] and sheathing[35] have appeared.

Other actual and potential applications for CM include impact compositions,[36,37] paints,[38] and as a modifier for rigid PVC.[39] Addition of 5–10% chlorinated polyethylene to tyre sidewall compounds has been found to significantly improve bond strength of the stocks.[40]

4. EPICHLORHYDRIN ELASTOMERS (CO: ECO)

4.1. Chemistry

Elastomers based on epichlorhydrin were announced in 1965.[41–43] Essentially saturated, high-molecular-weight, aliphatic polyethers with chloromethyl side chains, they are available as amorphous homopolymers of epichlorhydrin (CO) and as 1-to-1 mole copolymers of epichlorhydrin and ethylene oxide (ECO).

Basic structures and typical chemical compositions are shown in Table 4.

TABLE 4

STRUCTURES AND RAW POLYMER PROPERTIES OF EPICHLORHYDRIN ELASTOMERS

	Homopolymer (CO) $(-CH_2-CH-O-)_n$ $\vert$ CH_2Cl	Copolymer (ECO) $(-CH_2-CH-O-CH_2-CH_2-O-)_n$ $\vert$ CH_2Cl
Epichlorhydrin	100	65
Ethylene oxide	0	35
Chlorine content, %	38·4	25
Specific gravity	1·36	1·27

Preparative methods for experimental epichlorhydrin homopolymers, copolymers with propylene oxide, ethylene oxide and allyl glycidyl ether, and various terpolymers have been described.[44–46]

Patents have appeared in recent years claiming preparation of epihalohydrin polymers having molecular weights of 5000–300 000 by solution polymerisation with a catalyst comprising trialkyl aluminium, water and an ether.[47] Polymer variations studied have included grafting epoxides via an ester group, copolymerisation of epichlorhydrin and polymethylmethacrylate with BF_3–etherate catalyst,[48] and anionic copolymerisation of benzalacetone and epichlorhydrin.[49] Patents have also been granted for suspension condensation of epihalohydrins and polyalkylene polyamines[50] and for crosslinked polymers containing dicyclopentadiene rings.[51]

Possible future polymer modifications to provide pendent unsaturation may permit sulphur or sulphur-donor cures. This should shorten cure cycles without greatly affecting heat and fluid resistance.

4.2. Basic Properties

The two forms of epichlorhydrin elastomer have certain properties unmatched in the other.

The homopolymer, CO, has a brittle point of $-15°C$, excellent resistance to ozone and weathering and good resistance to heat and swelling in oils. Gas permeability is outstandingly low, being less than half that of butyl (IIR) rubber. Flame, ozone and heat resistances of the homopolymer are superior to those of the copolymer.

The copolymer, ECO, has a brittle point of $-40°C$, is much more resilient and is more suitable for low-temperature applications. Air permeability is comparable with medium acrylonitrile NBR.

Vulcanisates of both polymers exhibit low swell in water and aliphatic solvents. Tensile strength is typically 14 MPa, elongation 200–350%. Compression set at 100°C is similar to cadmium diethyl dithiocarbamate (Cadmate®)-cured NBR or post-cured ACM. Electrical properties and radiation resistance are considered to be relatively poor.

With compounds containing 40 phr filler and no plasticiser, volume change of CO and ECO after 70 h at 150°C in ASTM No. 3 oil has been reported as 11% and 15% respectively, compared with 24% and 8% for medium- and high-acrylonitrile NBR and 14% for ACM.[52] The same reference includes data on long-term air ageing at 150°C, immersion in hot oils and fuels, flex life, bond strength and physical properties at elevated temperatures. Long-term comparative air ageing, stress relaxation and property changes in ASTM No. 3 and commercial oils for red-lead-cured ECO, NBR, ACM, IIR, EPDM and FKM have been reported.[53]

4.3. Commercial Grades

There are two American sources offering grades as shown in Table 5. Epichlorhydrin elastomers are also produced in pilot-plant quantities in Japan.

4.4. Compounding and Vulcanisation

CO and ECO polymers process readily, having inherently good tack, low mill shrinkage, good extrusion characteristics and mould flow. Relationships between Mooney viscosity and heat built up under practical shear rates and temperatures, and the effects of compounding ingredients and cure systems on flow properties, have been studied.[54–56] Carbon black or conventional non-black fillers such as silica, silicates, treated clays or talc may be used as reinforcing agents or extenders. Ester-type plasticisers are compatible.

Vulcanisation may be accomplished with a variety of reagents that react difunctionally with the allylic chlorine group including diamines, urea, thioureas, 2-mercaptoimidazoline and ammonium salts.[45]

TABLE 5
AVAILABLE GRADES OF EPICHLORHYDRIN ELASTOMERS

Producer	Trade name	Grade	Type
Hercules (USA)	Herclor	H	Homopolymer
		C	Copolymer
BF Goodrich (USA)	Hydrin	100	Homopolymer
		200	Copolymer

Vulcanisate heat resistance is dependent upon the presence of an acid acceptor/cure activator, the stabiliser system and the curative selected. The effectiveness of common ingredients has been compared.[52] Lead based acid acceptors are by far the best. Inclusion of a stabiliser in ECO or CO increases vulcanisate heat resistance either by acting as an antioxidant or by repairing broken chains. Dibasic acids such as azelaic acid act in the latter manner in the presence of lead-based acid acceptors. Among curatives, thioureas impart much better heat resistance than diamines, although the latter cure faster and give lower compression set.

Studies on heat stability of various vulcanised epichlorhydrin elastomer compositions at 150°C suggest that hydrated silica usefully extends service life as compared with carbon blacks. Earlier studies indicated that excellent tensile properties at high silica loadings and high crosslink densities were apparently due to a favourable silica-polymer interaction.[57,58]

Studies on the vulcanisation of epichlorhydrin abound. Patents exist for systems to improve vulcanisate compression set and resistance to degradation in oil, water and brine at temperatures in excess of 120°C and pressures as high as 28 MPa. Specific sulphenamides improve processing safety.

4.5. Thermal Degradation

Several studies have been made on the mechanism of thermal degradation. Aspects include chemical stress relaxation and evolution of hydrogen chloride at elevated temperatures and the stabilising effects of carboxylic acids,[59] the deleterious effects of metal contamination[60,61] and the degree of crosslinking, gel fraction and volume of oxygen absorbed at 175°C.[62,63]

4.6. Applications

Published data suggest that in the USA up to 60% of epichlorhydrin elastomers are used in automotive and mechanical applications, which include control-system hose, tubing, diaphragms and moulded products. Hoses to SAE-type J30R1 have been made to meet the higher temperatures associated with emission control together with fuel resistance. Widespread use in oilfield and industrial hose is envisaged.

5. POLYACRYLIC ELASTOMERS (ACM)

5.1. Chemistry

Polyacrylate or acrylic elastomers comprise polymers with a preponderance of acrylic ester groups, generally derived from ethyl or butyl acrylate. They were developed in 1940 by the USA Department of Agriculture and the first commercial products were marketed under the Hycar® trade name in 1948.

The earliest ACM elastomers were vulcanised by strong polyfunctional bases; however, at an early stage in their development reactive groups were introduced into the chain to facilitate crosslinking. The first such group was 2-chloroethyl vinyl ether copolymerised with ethyl acrylate. This polymer was cured with active diamine systems. Problems associated with these systems included poor processing, limited bin stability, mould fouling and poor low-temperature properties. These problems led to the investigation of alternative reactive sites, which still continues. Comonomers evaluated include vinyl chloroacetate, allyl glycidyl ether, non-conjugated dienes and others. All of these copolymer systems need a vulcanising agent to be added to the compound but copolymerisation with an acrylamide permits self-curing ACM polymers to be obtained. These species are claimed to be stable up to 150°C and the cure rate can be increased by acidic accelerators.

Monomers bearing reactive sites usually represent about 5% of the polymer composition, the major part of the remainder being acrylic ester. Polymers based on ethyl acrylate have excellent oil resistance but poor low-temperature properties, whereas those based on butyl acrylate show improved low-temperature properties but inferior oil resistance. Thus, some commercial grades of ACM are based on compromise blends of these monomers, possibly with the addition of other acrylates such as alkoxy alkyl acrylate.[64–67]

® Hycar is a registered trade name of BF Goodrich (USA).

Representative structures are:

$$\left[\begin{array}{c}-CH_2-CH-\\ |\\ C=O\\ |\\ O\\ |\\ CH_2\\ |\\ CH_3\end{array}\right]_x + \left[\begin{array}{c}-CH_2-CH-\\ |\\ O\\ |\\ CH_2\\ |\\ CH_2\\ |\\ Cl\end{array}\right]_y \text{ or}$$

$$\left[\begin{array}{c}-CH_2-CH-\\ |\\ O\\ |\\ C=O\\ |\\ CH_2\\ |\\ Cl\end{array}\right]_y \text{ or } \left[\begin{array}{c}-CH_2-CH-\\ \diagdown\\ CH\\ |\diagdown O\\ CH_2\diagup\end{array}\right]_y$$

| Ethyl acrylate | Chloroethyl vinyl ether (amine cure) | Activated chlorine: vinyl chloroacetate (soap cure) | Epoxide (variety of curatives) |

There are many alternative copolymerisation routes to lower the brittle point of ACM at minimal cost to oil resistance. Some ACM types resist embrittlement down to $-40°C$.[68]

5.2. Commercial Sources and Grades

A list of USA and European suppliers of ACM and their principal polymer grades constitutes Table 6. ACM is also produced in Japan by Nippon Zeon, Takeda Chemical and Mitsui Toatsu.

5.3. Vulcanisation

Major advances in the technology of ACM have permitted elimination of active amine curing agents, initially by a combination of ammonium salts of carboxylic acid, ammonium benzoate and adipate. Advantages claimed for this system include improved processing and bin stability, shorter cure cycles, less mould fouling and improved compression set. Disadvantages include a tendency to cause pitting of high-carbon steel moulds.

TABLE 6
COMMERCIAL GRADES OF ACM ELASTOMER

Country	Manufacturer	Trade name	Grades	Suppliers' comments
Italy	Montedison	Elaprim AR	152	General purpose; good low-temperature flex
USA	American Cyanamid	Cyanacryl	153	General purpose
			R	High-temperature and hot-oil resistant
			L	Good low-temperature flex
			C	Best low temperature
USA	Thiokol	Thiacril	44	General purpose
			76	Good resistance to heat and shaft corrosion
USA	BF Goodrich	Hycar	4021	Heat resistance above 150°C
			4021–45	Low Mooney viscosity; improved processing
			4041	Use up to 205°C
			4041 CG	For cements and adhesives
			4042	Better low-temperature flex than 4021
			4043	Low temperature to −40°C; high temperature to +180°C
Canada	Polysar	Krynac	882	Good processing; versatile cure systems; low corrosion; good low-temperature flex
			882–C	As 882, better processing

Another improved vulcanisation system comprises mercaptoimidazoline and thiourea in the presence of red lead or dibasic lead phosphite.[69,70] This offered similar advantages but with less-dramatic improvements in resistance to mould fouling and compression set. It also tended to bloom.

Most recently, 'sulphur-soap' cure systems have been evolved consisting of small amounts of sulphur with alkali metal stearates. These are easy to control and are less temperature sensitive, since cure rate and properties may be altered by varying the relative concentrations of curatives.[71,72] Mould corrosion is low. Recent work has suggested that crosslinks produced by this system suffer thermal degradation during heat ageing, as opposed to the oxidative degradation predominant in amine-cured compounds. Most commercial ACM polymers announced since the discovery of the sulphur-soap cure system have been designed to use it.[73]

Another system of interest is the use of peroxides with non-chlorine-containing ACM elastomers.

5.4. Compounding and Processing

As with other elastomers, compounding of ACM is specific to processing and end-use requirements. Newer copolymers show markedly improved processability especially with regard to mill sticking. Compounds may be made on conventional open mills or in internal mixers.

A typical ACM compound should contain a reinforcing filler, usually carbon black. Acidic fillers should generally be avoided but stearic acid is often included as a process aid. The desirability of antioxidant addition depends on the elastomer structure. Heat resistance of newer ACM types containing epoxide or activated chlorine reactive sites may be optimised by adding 1–2 phr of antioxidant. Shaft seal compounds may contain fibrous fillers and/or graphite to reduce lip friction. Plasticisers, if used, must have low volatility to withstand the hot-air post-cure normally given and subsequent service at elevated temperatures.

Mixed ACM compounds are processed by conventional techniques, including extrusion and calendering. Compression moulding is the most common but transfer moulding and steam curing are also used. The newer copolymers with improved cure systems can also be injection moulded. Most acrylic elastomer mouldings are post-cured to produce the optimum properties, typically for 6 h at 185°C.

Bonding to metals during vulcanisation is accomplished using the same silane primers employed for silicone and FKM elastomers.

5.5. Properties

ACM elastomers are used primarily for their combined resistance to heat, oil and oil additives. They can withstand limited exposure to temperatures as high as 204°C, which places them well above NBR but below silicones and fluoroelastomers. The heat-resistant polymers can withstand 168 h in hot air at 175°C[73] with relatively little change, largely due to the saturated-polymer backbone which also imparts resistance to oil additives, especially sulphurised types for lubrication under extreme pressure conditions. Overall oil-swell resistance is similar to that of medium–high acrylonitrile NBR.[52] Copolymers based upon ethyl acrylate can withstand temperatures in excess of 180°C in contact with oils with good retention of physical properties and low volume swell. ACM has good inherent resistance to ozone, oxidation and aliphatic solvents but is more prone to hydrolysis than most other elastomers. Tensile strength values up to 14 MPa and elongation to 200% after post-cure are typical. Compression set of 40–60% after 70 h at 150°C has been reported for a soap-cured ACM.[74]

5.6. Applications

A survey has indicated that about 80% of USA consumption of ACM elastomers comprised seals, O-rings and gaskets; 10% was used in adhesives, sealants and coatings and the rest in miscellaneous items, probably largely mouldings. ACM is used in automotive transmission, valve stem and pinion seals. As pinion seals, it is among the few elastomers other than FKM able to withstand premature attack by sulphurised and other 'extreme pressure' oil additives. Softening in some EP lubricants may be a problem. The wider working temperature range of newer ACM types has helped extend their use to ignition-cable sheaths, spark-plug and dust boots, emission-control hose, belts, rolls, tank linings, sealants and adhesives. They are used as binders for cork or asbestos gasketing.

6. ETHYLENE ACRYLIC ELASTOMER

6.1. Chemistry

A new class of heat-resistant rubber bassed on a carboxylated copolymer of ethylene and methyl acrylate was announced in 1975 by Du Pont under the trade name Vamac®.[75] Indicated structural units are:

$$(-CH_2-CH_2-)_x \quad (-CH-CH_2-)_y \quad (-R-)_z$$
$$\qquad\qquad\qquad\qquad \underset{OCH_3}{\overset{C=O}{|}} \qquad \underset{OH}{\overset{C=O}{|}}$$

$$\text{Ethylene} \qquad \text{Methyl acrylate} \qquad \text{Cure site monomer}$$

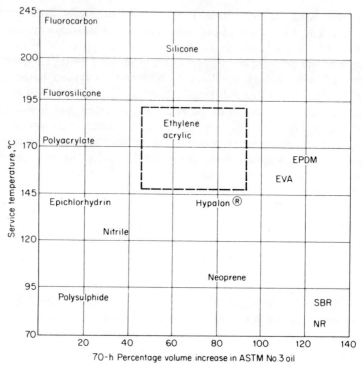

FIG. 1. Comparative heat and oil resistance of various elastomers.

6.2. Properties

Ethylene acrylic rubber was designed to fill an apparent gap in the property spectrum of heat- and oil- resistant rubbers (see Fig. 1). It fits into an area between Hypalon® (CSM) (see Section 2) and silicone rubber (see Section 7). The moderate oil resistance and useful service temperature range of $-30°C$ up to $+180°C$ indicates an ASTM D 2000 Class E designation.

In addition, this elastomer has complete immunity to ozone attack.

The properties for a typical ethylene acrylic seal compound[76] are shown in Table 7.

6.3. Compounding

6.3.1. Vulcanisation

Ethylene acrylic rubber can be vulcanised with primary diamines or peroxides.[77] Usually, methylene dianiline is used to provide safe pro-

TABLE 7

Original Properties	
Tensile, MPa	9.7
100% Modulus, MPa	6.2
% Elongation	227
Hardness	74
Specific gravity	1.38
Compression set (149°C)	
22 h, %	25.7
70 h, %	26.6
Low-temperature test per ASTM D 1053	
T-2	6.1°C
T-5	22.2°C
T-10	27.5°C
T-100	43°C
Low-impact test per ASTM D 746	
Pass	37.2°C
Fail	−40°C
Oven Aged	
70 h at 300°F (149°C)	
Tensile change, %	+12
100% Modulus change, %	+22
Elongation change, %	−19
Hardness change, points	+4
70 h at 350°F (177°C)	
Tensile change, %	+15
100% Modulus change, %	+51
Elongation change, %	−31
Hardness change, points	+7
Oil Aged—1 week at 300°F (149°C)	
ASTM No. 1 oil	
Tensile change, %	+23
100% Modulus change, %	+28
Elongation change, %	−7
Hardness change, points	+3
Volume change, %	−2
ASTM No. 3 oil	
Tensile change, %	−6
100% Modulus change, %	−18
Elongation change, %	+1
Hardness change, points	−26
Volume change, %	+38.3
Automatic transmission fluid	
Tensile change, %	+12
100% Modulus change, %	+1
Elongation change, %	+1
Hardness change, points	−8
Volume change, %	+12.6

cessing, relatively fast cures with good physical properties. More recently, a faster curing system based on hexamethylene diamine carbamate (Diak® 1)[78] has been developed.

Peroxide curing is normally restricted to non-black cable jacketing formulations because physical properties are not sacrificed. However, in black-filled compounds used for moulding, a peroxide cure will give lower physical properties. Hence an amine cure is preferred.

6.3.2. Fillers and Plasticisers
The polymer is normally supplied in the form of a masterbatch for ease of handling, either with 20 parts of SRF black (Vamac® B-124) or with 23 parts of fumed silica (Vamac® N-123). The former is used for general-purpose applications and the latter for cable insulation jackets.

Additional carbon black can be added to B-124, or additional silica or other white fillers to N-123.

Plasticisers are restricted to polyester types which provide a good balance between low-temperature performance and heat resistance.

A specific blend of process aids and mill release agents is mandatory to allow easy processing.

6.4. Applications

Potential applications for ethylene acrylic rubber lie in the area normally considered established for silicones, CSM, EVA and polyacrylates as indicated in Fig. 1.

Hence, automotive parts such as O-rings, shaft seals, spark-plug boots and other underbonnet miscellaneous parts can be considered and are now made in this elastomer. Radiator hose for diesel engines, where heat-, oil- and engine-coolant resistance are required, is also under development.

In the wire and cable industry, ethylene acrylic rubber is finding expanding use in ignition-wire sheathing, and for medium-voltage cables where its higher temperature rating and zero halogen content makes it an alternative to CSM. This is particularly important where a low toxicity and smoke-generation level is required to meet new, more-stringent flammability regulations such as for cables used in ships and submarines.

® Diak is a registered trade name of Du Pont (U.K.) Limited.

One property of ethylene acrylic rubber concerns its ability to absorb energy over a wide temperature range, as distinct from butyl rubber which peaks at about $-20°C$ (Fig. 2). Tan δ is the loss tangent which shows how much energy is absorbed by a system. The higher the number the better is the system's ability to dampen out vibrations. This leads to a further important application in high-temperature dampening applications such as gear-box and engine mounts which must operate outside the normal temperature range covered by natural rubber or SBR. The damping characteristics are also constant over a frequency range of 10-1000 Hz at a constant temperature.

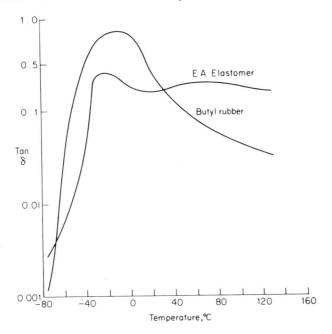

Fig. 2. Vibrational damping. Tan δ v. temperature.

7. SILICONE ELASTOMERS (MQ, PMQ, VMQ, PVMQ, FC)

7.1. Chemistry

Rapid growth of silicone elastomer consumption since World War II is attributable to a unique combination of properties relative to organic elastomers. This uniqueness results from molecular structure consisting of long chains of very strong, thermally stable silicon–oxygen ($-Si-O-$) bonds encased by organic groups. Compared to organic rubber polymer chains, silicone elastomers have large molar volumes and very low intermolecular attractive forces.[79]

Silicone polymers are made from organosilicon intermediates, or monomers, prepared from elemental silicon.[79] These monomers are compounds of the type SiR_nX_{4-n}, where R is an alkyl or aryl group and X is a group which can be hydrolysed to $-SiOH$, such as chlorine or alkoxy.

7.1.1. Heat-vulcanisable Polymers

Polydimethylsiloxane, ASTM D 1418 designation MQ, is the basic silicone polymer. It is a very-high-viscosity fluid or gum composed mainly of linear polydimethylsiloxane chains, produced by condensation polymerisation. Commercial gums usually contain between 3000 and 10 000 dimethylsiloxy units in the average chain:

$$(CH_3)_3SiO-\left[\begin{array}{c} CH_3 \\ | \\ Si-O \\ | \\ CH_3 \end{array}\right]- \quad n = 3000-10\,000$$

Substitution of vinyl groups ($-CH=CH_2$) for less than 0·5% of methyl groups yields polymers more responsive to peroxide vulcanisation agents. They require less peroxide for cure and show improved compression-set resistance. The vinyl group is neutralised during cure, giving a completely saturated polymer. Most commercial polymers today are copolymers containing vinyl groups. They bear the ASTM designation VMQ.

The good low-temperature flexibility of MQ rubber can be further improved by substitution of 5–10% phenyl ($-C_6H_5$) or ethyl ($-CH_2CH_3$) groups for some of the methyl groups in the polymer chain. Phenyl-rich copolymers provide elastomers resistant to radiation.[79] Phenyl-substituted silicone rubbers are designated PMQ.

MQ and VMQ rubbers are more resistant to swell in acetone and diesters than in aliphatic and aromatic hydrocarbons. The characteristics can be reversed by replacement of a methyl group on each silicon atom by a more polar entity, such as the trifluoropropyl group ($CH_2CH_2CF_3$). The resultant polymers are fluorosilicone elastomers (FVMQ) which are discussed in Section 8.

7.1.2. Room-temperature Vulcanising (RTV) Polymers

RTV compounds are based on low-molecular-weight silicone polymers with reactive and groups and a polyfunctional compound.

$$XO - \left[\begin{array}{c} CH_3 \\ | \\ Si-O \\ | \\ CH_3 \end{array} \right] - X \quad n = 100\text{--}1000$$

Base polymer

As in high-molecular-weight, heat-cured polymers, some methyl groups can be replaced by phenyl or trifluoropropyl groups to improve low-temperature flexibility and resistance to aromatic and aliphatic hydrocarbons respectively. End-group reactivity depends on the cure system; X is typically hydrogen, silanol, dimethylvinylsilyl, or alkyl of one to four carbon atoms. Systems have been developed that cure by reacting with atmospheric moisture.[80,81]

7.2. Commercial Sources and Grades

There are several manufacturers of silicone rubbers, most of whom supply raw polymers (gums), pre-compounded gums and RTV types. Other than phenyl- and vinyl-containing types, special grades include materials with enhanced tensile and tear strengths (tensile to 11 MPa), extreme high-temperature resistance (to $300 + °C$), improved insulation resistance for wire and cable applications, ultra-low compression set and flame resistance respectively. Also available are grades requiring no post-cure, 'self-bonding' types, others with low vapour pressures and outgassing for aerospace and vacuum systems, 'food contact' types and medical grades in both solid and RTV form suitable for implantation and other biomedical applications.

Major American producers are Dow Corning, General Electric, Union Carbide and Stauffer-Wacker. In Europe producers include Dow Corning, Bayer, and Wacker-Chemie (Germany) and Rhone-Poulenc (France). Silicone elastomers are also produced in Japan by Toshiba and Shin-Etsu, and in the USSR.

7.3. Vulcanisation Systems

Silicone elastomers may be cured in various ways:

1. by free-radical crosslinking;

2. by crosslinking of the attached vinyl or alkyl groups; through reactions with silylhydride groups.
3. by crosslinking linear or slightly branched siloxane chains having reactive end groups, such as silanol. This is a typical RTV system, yielding Si—O—Si crosslinks.

7.3.1. Heat-curable Compounds
Heat-cured silicone rubber compounds are normally cured in the presence of one or more organic peroxides such as benzoyl peroxide, tertiary butyl perbenzoate and bis(2,4-dichlorobenzoyl) peroxide.[82,83]

7.3.2. RTV Compounds
Many, if not all, curing systems for RTV silicones fit three general classifications:

1. Condensation cure—moisture independent. RTV products containing these curing systems will all cure in deep sections, independent of atmospheric moisture. They are known as 'two-package RTVs', because curing agent and/or catalyst must be added just prior to use.
2. Condensation cure—moisture dependent. Crosslinking occurs when the compound is exposed to atmospheric moisture, hence depends on the rate of diffusion of moisture into the rubber. Moisture-dependent RTV compounds are known as 'one-package RTVs', since the curing agent and catalyst are incorporated in the base compound during manufacture.[79,84]
3. Addition cure. Crosslinking occurs without formation of volatile products, and independent of the presence of air or atmospheric moisture.

7.4. Compounding
Silicone elastomer suppliers have available compounding information to meet over 100 military and industrial specifications. Much of this data is readily available.

7.4.1. Heat-curable Compounds
A typical silicone compound contains polymer, reinforcing and/or extending fillers, processing aids or softeners, special additives (blowing agents for sponge, etc.), pigments and one or more peroxide curing agents.

Polymer. Pure polymers may be used but it is usually easier and more economical to compound from reinforced gums. These are mixtures of pure gum, processing aids, and highly reinforcing silica and may or may not contain special additives.

Reinforcing fillers. Fume-process silicas reinforce silicone elastomers more than any other filler. Because this silica is of high purity, compounds containing it have excellent electrical insulation properties. Silica aerogels can be used for moderately high reinforcement, but they increase water absorption. Carbon blacks give only moderate reinforcement and retard cure with aryl peroxide curing agents. Their principal use is in electrically conductive or antistatic compounds.

Extending fillers. Ground silica and calcined china clay are used in compounds for mechanical and electrical goods. Calcined diatomaceous silica is used in electrical compounds, low-compression-set stocks, and mechanical goods stocks to reduce tack. Other extending fillers include calcium carbonate, zirconium silicate and zinc oxide.

Additives. Organic and some inorganic pigments adversely affect the heat ageing of silicone. Red iron oxide is suitable both as a pigment and a heat-ageing improver. Nitroso-based blowing agents may be used. Preparation and properties of sponge silicone stocks have been described.[85]

7.4.2. *RTV Compounds*
RTV rubbers are sold as 'ready to use' proprietary products requiring no further compounding.

7.5. Properties
Useful summaries of the properties of silicone elastomers are available.[79] A key attribute of silicone rubber is its ability to retain a large proportion of its initial properties after long periods of time over a wide range of temperatures. Also, initial physical properties tend to be relatively constant with varying temperatures; hence, while not as strong initially as some elastomers or as good a dielectric, the properties in service of silicone rubber tend to be more predictable.

Tensile strength is generally lower than for organic elastomers but 'high strength' types widen the range of values from 4·5 MPa to 11 MPa. Strength, as well as most other properties, depends on base polymer, fillers, special additives and state of cure. Elongation compares favourably with organic elastomers, ranging up to 900% in some cases. Tear strength varies from 8 kN m^{-1} to 35 kN m^{-1}. Again, these figures remain quite stable at high temperatures and under fatigue cycling. Hardness, as with most rubbers, can be varied considerably depending upon compounding techniques. Typical values range from 25 Shore A to 90 Shore A. Compression-set resistance is generally accepted as being better than that of any other elastomer, except bisphenol-cured FKM.[86] Values of 7–15% after 70 h at 150°C are not uncommon with special low-compression-set silicone grades.

Good stability during high-temperature heat ageing may not be obtained in the absence of air (oxygen) as under these conditions silicone and fluorosilicone elastomers can revert to a paste or powder at a lower temperature than expected, probably due to hydrolysis promoted by trapped moisture. Only slight ventilation is required to prevent reversion at temperatures up to 260°C.

'Useful life' for silicone at continuous elevated temperatures has been reported as 2 years at 150°C to 84 days at 260°C.[79]

At low temperatures, general-purpose grades resist embrittlement down to -75°C and retain flexibility at -50°C. Special grades are serviceable down to -115°C.

Correctly formulated silicone elastomers retain good electrical properties at temperature extremes under moisture and ageing conditions that severely affect many other insulations. They have excellent resistance to water, steam, weather and oxidation. Abrasion resistance of standard types is poor. After burning, they leave a non-conductive char. This is useful in certain types of cable.

Excellent chemical, solvent and oil resistance can be obtained, especially by using FVMQ polymers. General-purpose types show moderate swell in mineral and engine oils, rather higher swell in transmission fluids. They swell severely in petrol and diesel fuels, aromatic oils compounded for extreme pressure conditions, solvents and ketones. Oxidised lubricating oils can promote reversion.

7.5.1. Heat-curable Compounds

Representative physical properties of selected formulations are shown in Table 8. In view of the wide variety of grades offered, these figures should be taken as a basic guide only.

TABLE 8

TYPICAL PROPERTIES OF SELECTED HEAT-CURABLE SILICONE RUBBERS

Property	General purpose	Low compression set	High performance	Extreme low temperature	Electrical insulation
Hardness, Shore A	40–90	50–80	30–70	25–75	35–90
Tensile strength, MPa	4·1–8·9	5·5–7·5	8·2–9·3	5·5–8·3	6·4–11
Elongation, %	100–650	150–250	400–850	150–800	150–875
Tear strength, kN m^{-1} (ASTM D 624, Die B)	8·8–16·6	11–13·5	28–33	8·8–17·5	14–39
Temperature range, °C	−55 to +260	−55 to +250	−80 to +250	−115 to +260	−55 to +260
Compression set, %—22 h at 177°C	12–37	10–18	30–38	20–75	—
Heat-ageing changes—70 h at 200°C					
Tensile strength, %	−5		−10	−5	
	−15		−25	−20	
	−10		−25	−10	
	−30		−35	−40	
Elongation, %	+2 to +6		+5 to +8	+2 to +5	
Hardness, Shore A	+7 to +60	+7 to +15	+40 to +50	+65 to +150	
Volume change in ASTM No. 3 oil after 70 h at 80°C, %	1·9 × 10^{14}				0·1–2 × 10^{16}
Volume resistivity, ohm cm^{-1}	—	—	—	—	3·0–3·5
Dielectric constant, 60 cs^{-1}	—	—	—	—	0·004–0·02
Power factor, 60 cs^{-1}					

7.5.2. *RTV Compounds*

Physical properties of RTV compounds vary slightly depending upon the type of cure employed.

7.6. Applications

Applications for RTV grades depend partly on the cure system. One-package condensation cures are used for atmospheric crosslinked caulks and sealants; condensation-cure moisture-independent systems for elastic parts; coatings; adhesives; encapsulants; and therapeutic gels. Addition-cured RTV uses include flexible moulds, dip coating, potting and encapsulation of electrical components.

Solid silicone elastomers are used in the electrical industry as insulation for nuclear power and ships' cables, apparatus lead and electronic hook-up wire, and appliance and fixture wire. Other applications include capacitor bushings, coated glass sleeving, tubing and television corona shields.

In the automotive industry, applications include spark-plug boots, ignition-cable sheathing, heavy-duty radiator hose, valve-cover gaskets, and crankshaft and transmission seals. In the appliance industry, major applications for silicone elastomers are oven-door and washer-dryer gaskets.

Predictably, the wide operating temperature range of silicone elastomers finds uses in aerospace applications. Typical examples are air frame opening seals, pressure-regulator seals, wire insulation and hydraulic-system seals. Construction industry applications are predominantly RTV sealants. Miscellaneous applications include roll coverings, cellular sheeting and shapes.

8. FLUOROSILICONES (FVMQ)

8.1. Chemistry

Fluorosilicone elastomers combine in large measure the fluid and heat resistance of fluorocarbons and the low-temperature flexibility of silicone rubbers. Available commercially since 1958, they are made by ionic polymerisation.[87,88] The molecular weight is approximately 6000 but lower-molecular-weight products are available for sealant and fluid applications.

Uncompounded FVMQ gum, designated 'Silastic LS-420', is described as trifluoropropyl methyl polysiloxane, with a minor level of vinyl comonomer to facilitate peroxide vulcanisation.

More recently, hybrid copolymers derived from trifluoropropyl methyl polysiloxane and fluorocarbon segments have been studied to retard reversion in oxygen-free hot applications, and halogen-containing polymers for improved fire retardancy have been prepared. Reviews of this topic are available.[89]

8.2. Grades

The principal commercial polymer grades are shown in Table 9. One- and two-part RTV sealants based on FVMQ are also available.

TABLE 9

PRINCIPAL ELASTOMERIC FLUOROSILICONE GRADES

Designation	Description	Vulcanised specific gravity
Silastic LS-420	Polymer base	1·25
Silastic LS-422	84% base, 16% reinforcing silica filler	1·38
Silastic LS-53U	Compounded base for moulding	1·40
Silastic LS-63U	Compounded base for extrusion and calendering	1·45

8.3. Properties

As compared with fluorocarbons, FVMQ vulcanisates generally have lower hardness, modulus, elongation and tensile strength but at temperatures above 100°C both tensile and tear strengths of FVMQ exceed that of FKM.[89] High tear strength FVMQ has been developed. Compression set of FVMQ at 150°C is similar to or better than amine-cured FKM but not as good as bisphenol-cured fluoroelastomer or standard silicones, and FKM demonstrates superior resistance to long-term stress relaxation at 150°C in the absence of air.[90] Performance of FVMQ and FKM in static sealing applications over the temperature range −54°C to +150°C has also been compared with results favourable to FKM.

With a solubility parameter of 9·6, similar to that of FKM, the resistance of FVMQ to aliphatic and aromatic hydrocarbons is very good. Long-term hot-air resistance at 230°C may be obtained, combined with low-temperature stiffening at −55°C and brittleness at −66°C.[89] A TR-10 value of −60°C has been reported. Extended heat resistance of FVMQ is not as good as that of FKM but the combination of resistance to degradation in hot air and low-temperature flexibility is notable. Heat resistance may be enhanced by incorporation of aluminium or cadmium oxides. Dielectric properties exceed those of FKM.

8.4. Applications

Owing to very high polymer volume cost (about twice that of FKM), FVMQ is used only where its unique balance of properties is essential. Even so, 60% is thought to be used in industrial applications, compared with 40% in aerospace applications. Typical applications include O-rings, seals and flexible couplings for jet-engine fuel systems, fuel-pump diaphragms and carburettor valves, mountings for engines and electronic components, and electrical connectors to MIL-C-26500 specification. Diaphragms may be fabric-reinforced. RTV compositions are used as fuel-tank sealants in high performance aircraft. Development of a highly water-resistant moulding compound for electrical applications has been described.

9. FLUOROCARBONS (CFM AND FKM)

9.1. Chemistry

Fluorinated polymers date from the discovery of poly(tetrafluoroethylene) in 1938.[91] Highly fluorinated polymers are very stable and possess exceptional resistance to oxidation attack, chemicals, certain solvents, weathering and flame. This stability has been attributed to the high strength of the C–F bond as compared with the C–H bond, to steric hindrance and to strong van der Waals' forces.

Commercialisation of elastomers containing sufficient fluorine to impart high stability occurred in 1955 when copolymers of vinylidene fluoride VF_2 ($CH_2=CF_2$) and chlorotrifluoroethylene CTFE ($CFCl=CF_2$) were introduced. These copolymers, ASTM designated CFM, contain more than 50% combined fluorine and are available as KEL-F® in the USA and Voltalef® in France. Presence of chlorine in the polymer reduces heat resistance relative to co- and terpolymers containing fluorine substitution only.

Elastomeric copolymers of vinylidene fluoride and hexafluoro-

---

® KEL-F is a registered trade name of 3M(USA).
 Voltalef is a registered trade name of Ugine (France).

propylene HFP (CF_3—$CF\!=\!CF_2$) were announced in 1956 and details of polymerisation, curing and vulcanisate properties have been described.[92-94] They were commercialised as Viton® A types and subsequently Fluorel®. About 2 years later, a terpolymer of vinylidene fluoride, hexafluoropropylene and tetrafluoroethylene TFE ($CF_2\!=\!CF_2$) became available.[95] This contains a nominal 68% combined fluorine as compared with 65% for commercial VF_2/HFP copolymers, hence shows even better long-term resistance to heat, swelling in oils and solvents and chemical degradation, e.g. from certain oil additives. This polymer became known as Viton® B.

In the early 1960s copolymers of VF_2 and 1-hydropentafluoropropylene HPTFP (CF_3—$CF\!=\!CHF$) and terpolymers of these with TFE were disclosed.[96] Tecnoflon® S and T copolymers contain a nominal 62% fluorine. Both these and the preceding polymer groups bear the ASTM designation FKM.

More recently, a number of so-called tetrapolymers have been produced, so far of undisclosed chemical composition but built around the basic structure described above.[97,98] The essential difference between these and the foregoing fluorocarbon rubbers is the fact that they can be peroxide-cured (see also Section 9.2).

9.2. Commercial Sources and Grades

Known commercial sources and principal grades of FKM elastomers offered are listed in Table 10. These polymers are usually supplied additive-free although more recent types, notably the Viton® E series and B-910, the Fluorel® 217 series, the Tecnoflon® FOR series and DAI-EL® G-701 are sold containing proprietary vulcanisation systems.

It is interesting that compounds based upon FKM rubbers with viscosities normally considered high can, with reasonable care, be processed and moulded conventionally without use of volatile plasticisers. This suggests that this elastomer class possesses on unusual degree of thermoplasticity.

Special-purpose FKM types include Viton® C-10 and LM, low-viscosity grades for solution coatings and plasticising stiff fluoroelastomer compounds respectively; Fluorel® 2161 and 2171, process aid masterbatches for use with Fluorel® 2160 and 2170 respectively;

® Viton is a registered trade name of Du Pont (U.K.) Limited.
  Fluorel is a registered trade name of 3M(USA).
  Tecnoflon is a registered trade name of Montedison (Italy).
  DAI-EL is a registered trade name of Daiken Kogyo (Japan).

TABLE 10

PRINCIPAL GRADES OF FLUOROELASTOMERS COMMERCIALLY AVAILABLE

Country	Company	Grades	Probable monomer base
USA	Du Pont	Viton® A, A-HV, A-35, LM, C-10, E-60C, E-430, E-45, E-60	VF_2/HFP
		Viton® B, B-50, B-910, B-70	VF_2/HFP/TFE
		Viton® GH, GLT, VT-R-4590	Tetrapolymers, not disclosed
	3M	Fluorel® 2140, 2142, 2143, 2160, 2161, 2170, 2171, 2172, 2173, 2174, 2176, 2177, 2178, 2179, 2180, 2181	VF_2/HFP
	3M	KEL-F® 3700, 5500	VF_2/CTFE
		IF4	Perfluorobutyl acrylate
		LVS-76	Not disclosed
Italy	Montedison	Tecnoflon® SL, SH	VF_2/HPTFP
		Tecnoflon® T	VF_2/HPTFP/TFE
		Tecnoflon® FOR, FOR-LI, FOR-SR	VF_2/HFP
Japan	Daiken Kogyo	DAI-EL® G-501, G-601	VF_2/HFP/TFE
		DAI-EL® G-701	VF_2/HFP
France	Ugine Kuhlmann	Voltalef® 3700, 5500	VF_2/CTFE
USSR		SKF-26, 260, 260–1	VF_2/HFP
		SKF-32	VF_2/CTFE

Fluorel® 2172, an accelerator masterbatch for use in the Fluorel® 217 series; and Tecnoflon® FOR-SR, a curative-containing polymer designed for service in steam, hot water and acids.

More recent special-purpose grades include three peroxide-curable types. Viton® GLT is designed for improved low-temperature properties and is approximately 15°C better than Viton® A types and 12°C better than Viton® B-70. Viton® VT-R-4590 and Fluorel® LV-76 (bisphenol-cured) are designed for improved fluids resistance over conventional terpolymers of the B type. Finally, Viton® GH is suitable for continuous, low-pressure, hot-air curing and additionally has much improved steam and water and acid resistance.[99]

A developmental Viton® latex is also available.

9.3. Vulcanisation Systems

9.3.1. Diamine Cure Systems

Until the late 1960s, the vast majority of commercial FKM vulcanisates were prepared using derivatives of aliphatic diamines as crosslinking agents. Aliphatic diamines themselves are impractically reactive and scorchy, a deficiency overcome by using blocked diamines as their inner carbamates. Hexamethylene diamine carbamate, HMDA-C, frequently known commercially as Diak® 1, was the first such curing system and is still used today, although high-viscosity compounds containing it tend to be scorchy and exhibit limited uncured storage stability. Typical addition levels are 1–1·5 phr by weight. HMDA-C or peroxides such as benzoyl or *p*-chlorobenzoyl are the preferred crosslinking agents for CFM elastomers.

The most important blocked amine-type curing agent for FKM is biscinnamylidene hexamethylene diamine (DCND), known as Diak® 3. This provides the best balance of compound storage stability, processing safety, mould flow and cure-rate capability available in an amine system. Typical addition levels are 1·5–3·0 phr.

All practical blocked-amine-cured FKM compounds require the presence of a metal oxide to achieve commercially acceptable vulcanisates at practical press cure times. Also, an oven-post-cure of up to 24 h at 200°C in a vented oven is necessary to develop ultimate physical properties, especially compression set. A three-stage cure mechanism has been proposed.

1. Bases, e.g. metal oxide, react with polymer chains to form double bonds by elimination of HF.
2. Nucleophilic difunctional amine groups from decomposition of the curative react at the double bonds to form relatively unstable imine structures.

3. During post-cure, conjugated double-bond systems are formed by dehydrofluorination at points adjacent to double bonds formed in the first stage. These unsaturated centres then react to form additional heat-stable crosslinks. Water formed from the neutralisation of HF by the metal oxide is simultaneously removed. Thus, the post-cure cycle is necessary to remove centres of instability formed during the initial press (or steam) cure cycle.

Traditionally, 15 phr of low-surface activity magnesium oxide (MgO) has been found to give the best balance of compound storage stability, scorch and cure rate with blocked-amine curatives. For good resistance to water, aqueous chemicals and acids, up to 15 phr to litharge (PbO), should replace magnesia, although resistance to dry heat and compression set is somewhat impaired. A compromise is 10 phr each of zinc oxide and dibasic lead phosphite. This gives a rather lower rate and state of cure and set resistance, with greater processing safety and intermediate water and acid resistance. It is the favoured oxide system in CFM compounds for service in strong oxidising acids such as red fuming nitric acid or concentrated sulphuric acid.

Even with an optimum combination of polymer, amine curative and post-cure cycle, evaluators noted that in applications involving resistance to compression set and stress retention, such as O-rings, service life of fluoroelastomers was frequently limited not by inherent degradation resistance but by loss of sealing force and ability to recover after compression. Factory processing was marred by limited temperature and humidity-dependent compound storage life, and mould fouling due to reaction of amine with the metal surface.

These problems led to the development of FKM curing systems based upon aromatic polyhydroxy compounds. To be practical these require the presence of strongly basic organic adjuvents. Most recent systems are based on bisphenols. These latter provide outstanding resistance to compression set and stress decay at temperatures to 200°C, excellent compound storage stability and processing safety and dramatically reduced tendency to mould fouling.[100]

A 1973 paper has discussed practical aspects and potential mechanisms of FKM crosslinking systems in some detail.[101]

9.3.2. Bisphenol Cure Systems

Bisphenol cure systems are supplied ready mixed in the base polymers or separately as masterbatches.[102] System variations permit a wide range of cure rate and state and demoulding hot tear strength. Commercial examples of curative-containing types include Viton® E-60C, E-430 and B-910, Fluorel® 2170, 2173, 2174, 2176, Tecnoflon® FOR, FOR-LI and FOR-SR and DAI-EL® G-701. Fluorel® 2172 is an accelerator masterbatch used to adjust cure rates of the 217 series and Viton® curative masterbatches are supplied for incorporation into Viton® E-60, E-45 or all existing A and B types if required.[102]

Depending on the ratio of accelerator to bisphenol used, it is possible to obtain a virtually 'ideal' rheometer cure curve as shown in Fig. 3. The bisphenol curative provides extreme safety at processing temperatures, as shown by Mooney scorch. At 180°C a controlled delay allows good mould flow even in complex tools, followed by extremely rapid modulus development to a stable level.

Bisphenol cure systems also require the presence of a metal oxide, usually 3–5 phr of highly active magnesium oxide rather than higher levels of low-activity magnesia. The latter may, however, be used up to 17 phr in special compounds for enhanced bonding to metals. In addition, 1·5–6 phr or more of calcium hydroxide is necessary to act as a

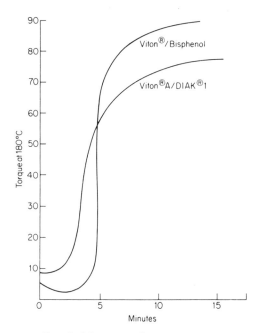

FIG. 3. Monsanto rheometer cure rate.

vulcanisation promoter, possibly by the controlled release of water. Length of the cure-induction time at moulding temperatures and safety at processing temperatures are inversely dependent upon calcium hydroxide concentration. Air oven post-cure at temperatures in the range of 230–260°C is normal with these cure systems, although compression set without post-cure is equivalent to post-cured, diamine-based systems.

Versatility of the newer vulcanisation systems has enabled suppliers to offer special grades with outstanding compression-set and stress-relaxation resistance for O-ring manufacture. Typical of these are Viton® E-60C, Fluorel® 2170 and 2174 and DAI-EL® G-701. Compression set of fluroelastomers versus polymer viscosity and cure system has been compared in suppliers' literature and elsewhere,[100,101,103,104] and is summarised in Table 11.

Long-term testing up to 9000 h at 150°C and 1200 h at 200°C in air and service fluids has confirmed that bisphenol-based cure systems have dramatically improved the compression-set and stress-relaxation resistance of FKM rubbers over their feasible working temperature range and has broadened that range.[90]

TABLE 11

COMPRESSION SET OF FLUOROELASTOMERS WITH VARIOUS CURE SYSTEMS

Polymer	Cure system type	Polymer ML at 100°C	O-Ring compression set, %: ASTM D 395, Method 'B'—70 h at 200°C in air
Viton® A	HMDA-C	65	45–50
Viton® A	DCND	65	60
Viton® A-HV	HMDA-C	160 (at 121°C)	35
Fluorel® 2160	Hydroquinone	60	26
Viton® E-60	Hydroquinone	60	25
Tecnoflon® FOR	Bisphenol	ca. 90	16–20
Viton® E-60C	Bisphenol	60	15–18
Fluorel® 2170, 2173, 2174	Bisphenol	60	15–18

9.3.3. Peroxide Cure Systems
These systems, applicable only to Viton® polymers GH, GLT and VT-R-4590, do not provide the same resistance to compression set and long-term stress decay (although better than diamine systems).

However, their main attribute is to provide compounds capable of being cured in low-pressure systems such as hot air or LCM without sponging or fissuring. In addition, they provide much improved resistance to hot aqueous media, e.g. steam, hot water and acids, and allow FKM elastomers to compete in the field normally held by EPDM or butyl elastomers.

To summarise this section, Table 12 rates the main property differences between the three types of curing system.

TABLE 12

COMPARISON OF FKM CURE SYSTEMS

Property	Blocked diamine	Bisphenol	Peroxide
Pressureless curing	Not applicable	Not applicable	1
Demoulding tear strength	1	1–2	3
Stress retention, 200°C air	3	1	2
Stress retention, steam/hot water	3	2	1
Acid resistance	3	2	1
Flex life	1	2	3

Rating: 1 better than 2 better than 3

9.4. Compounding

Probably owing to the simplicity of fluoroelastomer-based compounds, relatively little is published on compounding other than in polymer suppliers' technical literature. However, two useful summaries are available.[104,105]

The composition of a typical FKM compound is given in Table 13.

TABLE 13

	Parts by weight
Elastomer	100
Metal oxide	3–20
Cure promoter, e.g. calcium hydroxide (bisphenol cure)	1·5–6
Filler(s) (normally MT black)	5–50
Curing agents (if not in raw polymer)	1–3
Processing aid	0·5–3
Coagent (if peroxide-cured)	2

9.4.1. Fillers

Serviceability of FKM depends more upon inherent resistance to severe environmental conditions than upon physical properties. Also, high polymer cost dictates use of the maximum filler volume consistent with performance requirements. Reinforcing structure blacks increase tensile and tear strengths but impair processing of higher-viscosity polymers and disproportionately increase vulcanisate hardness; hence compounds containing them have higher polymer contents than those formulated with non-reinforcing thermal blacks or mineral fillers. Filler loadings and scope for compound cost reduction are severely restricted by the absence of effective plasticisers (see Section 9.4.2) and high gum stock IRHD values.[50-55]

9.4.2. Plasticisers and Process Aids

No fully compatible plasticiser for FKM, non-fugitive during post-cure and high-temperature service, has been found; hence polymer extension and improvement of low-temperature flexibility are largely impractical. If required for processing, lower compound viscosity is normally achieved either by polymer selection or by incorporation of 5–20 phr of a very-low-molecular-weight fluoroelastomer such as Viton® LM.

Process aids, typically low-molecular-weight polyethylenes or hard waxes such as carnauba, may be used at 0·5–3 phr, depending on type, to aid roll and mould release and to reduce die drag during extrusion.

9.5. Properties

Again, suppliers' literature is the most detailed source of information, although two survey articles[104,105] provide a useful outline.

The practical IRHD hardness range for FKM vulcanisates is 50–95. Most fall in the 70–85 region. Unlike bisphenol-cured vulcanisates (some of which are more stable), hardness of diamine-cured compounds may drop 10 points or more with temperature rise to 200°C.

Continuous potential service life of FKM in air has been estimated and is shown in Table 14.

TABLE 14

Temperature, °C	Hours service
230	> 3000
260	1000
290	240
315	48

Thermal degradation and specifically elimination of HF on heating FKM has been widely studied.[106,107] CFM is not as heat-resistant as FKM.

Compression-set resistance has been dealt with under Section 9.3 in connection with vulcanisation systems. However, Fig. 4 indicates the clear superiority of an FKM elastomer over other oil- and heat-resistant elastomers in a test designed to show the sealing-force retention of an O-ring at 150°C.

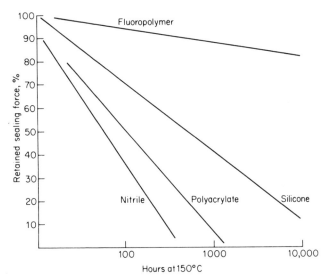

FIG. 4. Sealing-force retention of fluorocarbon versus other elastomers used for seals.

9.5.1. Fluids Resistance

FKM vulcanisates are highly resistant to aliphatic and aromatic hydrocarbons, chlorinated solvents and petroleum fluids. They show high swell in ketones, monoesters, ethers and certain proprietary fluids such as Skydrol® 500B, an alkyl aryl phosphate ester aircraft non-flammable hydraulic fluid, but they resist aryl phosphate ester types. Resistance to mineral acids can be good if suitably compounded but CFM is better than FKM in resisting strongly oxidising acids and alkalis. Unless compounded with litharge, resistance of FKM to hot water, steam and wet chlorine is relatively poor (but see peroxide cures—Section 9.3.3). Resistance to oil-well 'sour gas' is poor except for Viton® VT-R-4590 (see below).

® Skydrol is a registered trade name of Monsanto Corp. (USA).

Among chemicals, ammonia and amines tend to embrittle FKM, including amines present as antioxidants in some lubricants. Hot anhydrous HF and chlorosulphonic acid attack it.

One of the new peroxide-cured FKM elastomers has resulted in an improvement in the fluid resistance of these rubbers. Some typical comparisons with a normal B-type FKM elastomer are shown in Table 15.

TABLE 15

FKM POLYMER WITH ENHANCED FLUIDS RESISTANCE

Fluid	Test	Volume change, %	
		Viton® B	VT-R-4590
Methanol	7 days/24°C	+22	+4
Benzene	7 days/24°C	+15	+7
Jet-engine oil	7 days/200°C	+15	+10
Skydrol® 500B	7 days/121°C	+130	+42
70% Nitric acid	7 days/70°C	+21	+8

9.5.2. Low-temperature Properties

FKM parts rarely fail in service due to brittle fracture at low temperatures. However, operation of sensitive equipment such as fuel-metering systems, regulating valves, etc., can be impaired by stiffening of seals, diaphragms, etc., at sub-zero temperatures. Frequently, only a few degrees improvement in polymer low-temperature flexibility makes the difference between satisfactory and unsatisfactory field performance.

Two new FKM elastomers provide enhanced low-temperature flexibility without use of plasticisers or significant loss of fundamental polymer properties. One, Viton® B-70, is vulcanised with bisphenol or blocked polyamine systems; the other, Viton® GLT, is crosslinked with peroxides.[108,109] The data in Table 16 illustrate the performance of these two elastomers compared with more-conventional VF_2/HFP and VF_2/HFP/TFE polymers (A or B types).

It is worthy of note that successful sealing of aviation fuel down to $-54°C$ has been reported by an aircraft-fuel-system manufacturer using O-rings based upon a conventional bisphenol-cured E-type fluoroelastomer.[110] This indicates that where static seals are involved, conventional laboratory low-temperature tests may be poor indicators of actual field performance in that they tend to underrate the elastomer's true capabilities.

TABLE 16

FKM POLYMERS WITH ENHANCED LOW-TEMPERATURE PROPERTIES

	Viton® E-60C	Viton® B	Viton® B-70	Viton® GLT
O-ring set, 70 h at 200°C, %	18	27	26	30
Clash and Berg stiffness, °C	−17	−13	−21	−30
Brittle point, °C	−35	−41	−46	−51
Volume change in ASTM reference fuel C, 70 h at 100°C, %	+20	+18	+22	+23

9.6. Applications

Fluoroelastomers were originally developed to provide seals in aerospace applications subject to severe temperature and fluid exposure. A recent survey[111] indicates that approximately 75% of USA fluorocarbon rubber consumption is accounted for by O-rings, packings and gaskets. Automotive and other mechanical goods account for 12%, adhesives, sealants and coatings 5%, coated fabrics, sheeting and hose 4% each, and miscellaneous uses 3%. Applications in passenger cars are restricted by high polymer cost but fluoroelastomers are used in valve-stem seals, heavy-duty automatic transmission and pinion seals for high heat and oil additive resistance, and also in crankshaft seals and cylinder-liner O-rings for diesel engines. The service temperature range for FKM rotary shaft seals has been noted as −45°C to +200°C. They are increasingly used where downtime and replacement costs are high and to provide greater reliability where ACM or even NBR could be considered.[112] References can also be made to a comprehensive review of elastomeric shaft seals and materials.[113]

Resistance of FKM to hot corrosive gases should lead to new applications in automotive and industrial emission-control systems. Miscellaneous applications include roll and conveyor-belt covers for hot materials, oil and chemical valve linings, e.g. for tanker discharge systems, acid- and oil-resistant hose liners, pump impellers, cable sheathing for extreme conditions and flue-duct expansion joints.

10. PERFLUORINATED ELASTOMER (PFE)

Commercialisation in the mid-1970s of rubbers based essentially upon copolymers of tetrafluoroethylene and perfluoromethyl vinyl ether provided elastomeric materials without hydrogen atoms in the molecule, thereby dramatically increasing resistance to thermal, oxidative, chemical and aggressive fluid attack, even as compared with FKM.[29] Chemical resistance of PFE approaches that of PTFE, hence is almost universal.

10.1 Chemistry
The basic structure of PFE, known commercially as Kalrez®, is shown below:

$$\left[-(CF_2-CF_2)_x-(CF_2-CF)_y- \atop \qquad\qquad\qquad\quad | \atop \qquad\qquad\qquad\quad O \atop \qquad\qquad\qquad\quad | \atop \qquad\qquad\qquad\quad CF_3 \right]_n$$

TFE (60)   PMVE (40)

(Polymer contains $< 2\%$ of a fluorinated third monomer to permit vulcanisation.)

10.2. Availability
Owing to unusual processing requirements, parts based upon Kalrez® (PFE) are only available from the polymer supplier and agents. Special parts other than O-rings, sheet and card can be made but there are restrictions on diameter and thickness of mouldings. At present, metal-bonded dynamic shaft seals are not offered.

10.3. Properties
Salient advantages of PFE over FKM include the following:[114]

1. Continuous dry-heat resistance to 260°C, intermittent to 315°C.
2. Essentially unaffected by hydrocarbon and polar liquids.
3. Resistant to the majority of chemicals that attack other elastomers, including FKM, e.g. acrylonitrile, amines, ketones, styrene, vinyl chloride, hot sodium hydroxide, fuming nitric acid.
4. Resistant to oil-well 'sour gas'.

® Kalrez is a registered trade name of Du Pont (U.K.) Limited.

SPECIAL-PURPOSE ELASTOMERS

5. Resistant to high-temperature steam.
6. Very good long-term, high-temperature compression-set resistance.
7. Resistant to outgassing under vacuum to 300°C.[14]

Disadvantages of PFE include:

1. Slow recovery from compression at moderate temperatures, e.g. below 150°C.
2. Relatively high thermal coefficient of expansion, sometimes necessitating redesign of seal housings.
3. Limitations on size and complexity of finished parts.
4. High unit part cost.

PFE compounds are available covering the 70–97 Shore A hardness range. The highest hardness compound is designed for maximum seal-extrusion resistance under extreme service pressure and temperature conditions, e.g. 100+ MPa at 200+ °C.

Long-term heat ageing of a 90 Shore A PFE compound is demonstrated in Table 17. This should be compared with results for FKM (see Section 9), which has almost complete loss of elasticity after an average 125 days at 250°C or an estimated 60 days at 260°C.

TABLE 17

DRY HEAT AGEING RESISTANCE OF A 90 SHORE A PFE

Property	Original	500 days at 230°C	112 days at 260°C	28 days at 290°C	7 days at 315°C
Tensile strength, MPa	18·3	16·8	16·8	11·2	14·0
Elongation, %	120	220	240	320	230
Hardness, Shore A	90	87	87	86	87

The effect of temperatures from 200°C to 260°C on long-term O-ring stress-relaxation or sealing-force retention of PFE versus E-type FKM is shown in Fig. 5. It is notable that PFE rings at 260°C and FKM rings at 230°C reach zero sealing force after a similar period (40–50 days).

Table 18 compares the resistance of perfluoroelastomer and FKM to selected fluids and chemicals.

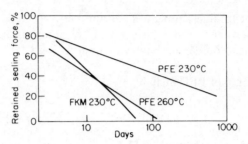

Fig. 5. PFE versus FKM: effect of temperature on O-ring sealing-force retention.

TABLE 18

CHEMICAL RESISTANCE; PFE VERSUS FKM (7 DAYS AT 24°C)

Medium	Volume increase, %	
	PFE	FKM
Benzene	3	22
Ethyl acetate	3	280
Methyl ethyl ketone	< 1	240
Tetrahydrofuran	< 1	200
Nitrobenzene	1	24
Nitrogen tetroxide	Satisfactory	280
Ammonia, anhydrous	Satisfactory	Embrittled

TABLE 19

PFE PARTS IN CHEMICAL SERVICE

Medium	Temperature, °C	Time, days
Acrylonitrile	Ambient	14
Methyl methacrylate	160	180
Mixed hydrazine/ammonia/sulphides	66–93	240
m-Phenylene diamine	230	180
Process steam	250	365
'Sour gas'	to 150°C/50 MPa	750
Styrene	Ambient	95
Sulphuric acid, 30% oleum	50	270
Vinyl chloride	Ambient	Satisfactory

Finally, Table 19 lists representative media and conditions where PFE seals function reliably but all other elastomers tested have failed.

10.4. Applications

The following industrial applications for Kalrez® have been noted:
1. Pressure-switch seals and O-rings in hot mercury/caustic mixture lines to 120°C.
2. Compressor O-ring seals in carbon dioxide at 230–245°C.
3. O-rings in mechanical seals in heat transfer fluid at 245°C.
4. A similar application in maleic anhydride at 170°C.
5. Oilfield deep well tube-to-packer V-ring seals in 14% sour gas for 2 years at 150°C. Predicted maximum service temperature is 200°C.
6. Protective electrical connector sleeves in geothermal 'logging' tools at operating depths to 2150 m, live steam to 260°C.
7. Seals in approximately 100 American chemical plants and 20 oil refineries.
8. Seals in chemical process equipment from approximately 25 American manufacturers.
9. Seals and septa in process and analytical instruments from nine American manufacturers.

11. ETHYLENE–VINYL ACETATE

11.1. Chemistry and Availability

Copolymers of ethylene and vinyl acetate, colloquially termed EVA, are elastomeric in the 40–60% VA range but commercial grades usually contain 40–45%. The copolymer, structural formula as shown below, is substantially amorphous at 40% VA content but retains some of the thermoplasticity of polyethylene.[115,116]

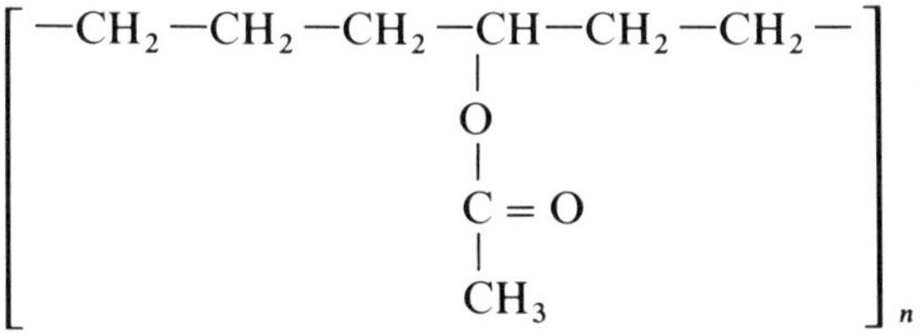

$$\begin{bmatrix} -CH_2-CH_2-CH_2-CH-CH_2-CH_2- \\ | \\ O \\ | \\ C=O \\ | \\ CH_3 \end{bmatrix}_n$$

Ethylene–vinyl acetate

Commercial grades of EVA elastomer are marketed as Levapren® and Vynathene®, variations being confined to VA content (40–60%) and viscosity, which is low (12–20 Mooney at 100°C).

11.2. Properties

Ethylene–vinyl acetate is characterised by good dry-heat, ozone and weather resistance. Heat resistance is better than EPDM but poorer than FKM and silicones, and may be ranked similar to ACM. Resistance to swelling in oil and solvents is less notable, being inferior to polychloroprene (CR) and NBR. Comparative oil resistance and low-temperature stiffening of EVA, CR, NBR, ACR, FKM and MVMQ has been reported.[117] ASTM D 746 brittleness values below -50°C may be obtained without plasticisers.

Vulcanised EVA copolymers can withstand 120°C continuously, or 140–150°C with a useful life up to 1 year. At 180–200°C, life can be measured in weeks. Degradation studies relating VA content and oxygen absorption have indicated that 40% VA is preferable to lesser levels, but that very low levels in the non-elastomeric range give best heat resistance.

11.3. Vulcanisation and Compounding

Compounding of EVA is basically simple. Cure systems are peroxide-based, usually in conjunction with a coagent such as triallyl cyanurate for increased state of cure. Peroxide selection depends on desired cure time, temperature, and freedom from odour. High-energy-radiation curing may also be employed.

Carbon black is the preferred filler for EVA, reinforcing types giving best physical properties. Mineral fillers and plasticisers should be selected carefully to avoid interference with vulcanisation. Low-volatility paraffinic oils and esters are satisfactory. For optimum heat resistance a polycarbodimide-type antioxidant may be used. Physical properties fall within the normal rubber range, with tensile strengths up to 25 MPa and elongations of 200–600%. Hardness ranges from Shore 60–85. Compression-set resistance over the temperature range 100–180°C is claimed to be superior to that of silicone elastomers. A post-cure may be given to remove peroxide decomposition products.

® Levapren is a registered trade name of Bayer AG, Germany.
  Vynathene is a registered trade name of United States Industries.

1.4. Applications

The heat resistance of EVA vulcanisates is utilised in moulded and extruded seals for use in heating installations and automotive applications, and also in cable insulations. Compounding principles and properties for the latter are established and compositions satisfying Underwriters Laboratory Specification UL 44 are practicable.[118] A comprehensive review of EVA polymers has been published.[119]

12. POLYSULPHIDE RUBBERS (T)

Introduced in 1939 by the Thiokol Corp., these rubbers still provide the widest range of solvent resistance available, although other elastomers may provide better specific resistance to certain classes of solvent (e.g. FKM). See also Fig. 1, Section 6.2.

12.1. Chemistry

Polysulphide rubbers are produced by reacting an organic dichloride and a polysulphide.

$$Na_2S_4 + Cl \cdot CH_2-CH_2 \cdot Cl \longrightarrow (CH_2CH_2S_4)_n + 2NaCl$$

Two distinct classes of product are manufactured as determined by the polymer end-groups.

12.1.1. Hydroxyl-terminated

Types A and FA are hydroxyl-terminated and can be designated by the structure $(-SS-R'-SS-R''-SS-)_n$. They are high-molecular-weight polymers and have to be plasticised chemically with a thiazole-type rubber accelerators such as MBTS to afford processability.

$-SS-R'-SS-R''-SS-)_n + R'''-SS-R'''$
  Polymer   Accelerator
  $\longrightarrow (SS-R'-SS-R''') + (R'''-SS-R''-SS-)$
                    Softened polymer

Redistribution of the disulphide/accelerator linkages occurs. These plasticised compounds can then be vulcanised with zinc oxide. Unfortunately, due to this method of plasticisation this class of polymer has poor compression-set resistance.

12.1.2. Mercaptan-terminated

Type ST is mercaptan-terminated and the structure is different since the polymer is depolymerised during manufacture.[120]

$$R-SS-R + NaSH \longrightarrow R-SSNa + RSH$$
$$R-SS-Na + NaHSO_3 \longrightarrow RSH + Na_2S_2O_3$$

TABLE 20

COMPARISON OF SOLVENT RESISTANCE OF THREE STANDARD POLYSULPHIDE RUBBER COMPOUNDS

Solvent	Volume swell after 1 month at 25°C, %[a]		
	A	FA	ST
Benzene	30	96	114
Toluene	24	55	79
Xylene	17	31	39
Carbon tetrachloride	20	36	48
SR-6[b]	6	10	10
SR-10[c]	−2	1	1
Glacial acetic acid	10	21	18
Butyl acetate	17	17	35
Dibutyl phthalate	6	7	8
Linseed oil	0	0	0
Ethyl alcohol	0	2	5
Ethyl glycol	0	1	3
Glycerol	0	2	1
Methyl ethyl ketone (MEK)	34	28	49
Methyl isobutyl ketone (MIK)	24	13	25
10% HCl	2	2	2
100% HCl	D	D	D
20% NaOH	1	2	2
10% HNO_3	D	D	D
10% H_2SO_4	2	2	2
50% H_2SO_4	D	D	D
Water	3	5	5

D = Decomposed.
[a] ASTM designation D 471-64T 'Change in Properties of Elastomeric Vulcanisate Resulting from Immersion in Liquids'.
[b] 40% Aromatic, blended reference fuel—60% isooctane, 20% xylene, 15% toluene, 5% benzene.
[c] 100% isooctane.

The Mooney viscosity can be adjusted during manufacture to the desired range since splitting of the disulphide groups occurs along the chain.

Vulcanisation can be subsequently carried out with oxodising agents such as metallic oxides, peroxides, quinoid compounds, etc.

By continuing the reduction in molecular weight, liquid polymers can be made from this class and are designated LP grades.

12.2. Properties

The important properties of polysulphide rubbers are solvent resistance and vapour impermeability. Typical solvent-, acid- and alkali-resistant properties are shown in Table 20, and permeability is compared with other elastomers in Tables 21 and 22.

Generally, Type A has best solvent resistance followed by FA and ST. Properties of the LP grades after cure are similar to the ST solid types.

TABLE 21

COMPARISONS OF THIOKOL ST WITH STANDARD COMPOSITIONS OF VARIOUS SYNTHETICS IN SPECIFIC PERMEABILITY[a] IN SR-6

Temperature	ST	SBR	Low nitrile	High nitrile	Chloroprene
24°C	0·12	13·4	2·7	1·1	3·5
38°C	0·19	15·4	3·5	1·3	4·1
50°C	0·26	20·1	5·0	1·5	5·0
65°C	0·37	25	7·4	7·4	7·3
80°C	0·53	27	10·7	2·5	10·7

[a] Specific permeability (g)(cm)/(24 h)(m²)

TABLE 22

SPECIFIC PERMEABILITY[a] OF THIOKOL ST IN VARIOUS SOLVENTS AT VARIOUS TEMPERATURES

Temperature	Methyl alcohol	Carbon tetrachloride	Ethyl acetate	SR-6	Benzene	Diisobutylene
24°C	0·03	0·28	1	0·12	3·6	0
38°C	0·07	0·43	1·3	0·19	4·6	0
50°C	0·16	0·7	1·8	0·26	6·1	0
65°C	0·36	1·07	2·5	0·37	8·7	0
80°C	0·94	1·4	3·4	0·53	12·0	0·006

[a] Specific permeability (g)(cm)/(24 h)(m²)

Some additional physical properties are given in Table 23. Note that compression-set resistance and low-temperature flexibility improve as one moves from the A to FA to ST grades. Heat resistance varies but is normally confined to 100°C continuous use and intermittent up to 150°C. The ST type is preferred. Some data have been reported[120] for oxygen-pressure resistance at 70°C where polysulphide rubber was unaffected.

TABLE 23

PHYSICAL PROPERTIES OF VARIOUS POLYSULPHIDE RUBBER COMPOUNDS

Polysulphide type	Type A	Type FA	Type ST
Tensile, MPa	5·4	8·6	8·6
100% Modulus, MPa	2·4	3·1	3·1
200% Modulus, MPa	5·1	5·5	8·3
Elongation, %	370	400	310
Hardness, Shore A	78	71	70
Compression-set resistance	poor	poor	satisfactory
Gas formation on milling	yes	no	no
Flexibility limit	−18°C	−45°C	−50°C

12.3. Compounding

Typical compounds would contain 60 parts of SRF black but utilise different curing systems dependent on type of polymer, as explained previously.

A Type FA polymer requires approximately 10 parts of zinc oxide plus 0·3 MBTS and 0·1 DPG as chemical plasticisers. A Type ST polymer can be vulcanised with the following combinations:

1. 1·5 parts p-quinone dioxime,
   0·5 parts zinc oxide,
   0·5–3 parts stearic acid;
2. 1 part p-quinone dioxime,
   10 parts zinc chromate,
   1 part stearic acid;
3. 6 parts zinc peroxide,
   1 part stearic acid.

Zinc peroxide is used for non-staining compounds. A liquid polymer (LP) can be used up to 5% in order to soften a compound based on the ST type only, and will co-cure.

12.4. Applications

The properties of polysulphide rubbers have established their use in fuel hose and tubing, binders for cork, putties and sealants, roll coverings, O-rings, gaskets and diaphragms. The use of these rubbers could well have been greater if it were not for the smell of fumes produced during processing.

The ST grade is the only suitable polymer for O-rings and seals because it is the only type that exhibits satisfactory compression set.

The liquid polymers are used, as expected, mainly as sealants and for impregnation or encapsulation. LP 3 and 33 are low molecular weight; LP 2, 12 and 32 are medium; and LP 31 high. LP 2 can be converted to a solid rubber, without shrinkage, at room temperature, by the addition of 7·5 phr or more lead dioxide.

Liquid polymers can also be dissolved in solvents such as dioxane or cyclohexane to produce solutions.

13. PHOSPHONITRILIC FLUOROELASTOMERS

Based on the inorganic rubber polydichlorophosphazene, pendent chlorine groups are replaced by partially fluorinated alkoxy species to form a hydrolysis-resistant fluoroelastomer. Preparation was first reported in 1968.[121]

A polymer with very high molecular weight (greater than 3 million) and broad molecular-weight distribution is available semi-commercially as PNF 200 and in black masterbatch form as PNF 230. Uncompounded polymer cost is high.

$$\left[\begin{array}{c} Cl \\ | \\ -P=N- \\ | \\ Cl \end{array}\right] \longrightarrow \left[\begin{array}{c} OCH_2CF_3 \\ | \\ -P=N- \\ | \\ OCH_2(CF_2)_3CF_2H \end{array}\right]_n$$

Inorganic rubber   PNF

Processing characteristics of the polymer reflect its abnormally low 100°C Mooney viscosity of 14. PNF is solvent-resistant with a potentially wide operating temperature range (-50°C to $+200$°C). T_g is -68°C, hence low-temperature properties similar to fluorosilicone may be expected in vulcanisates. Heat and fluid resistance and low-temperature properties of PNF have been compared with other oil-resistant elastomers.[122]

PNF may be conventionally compounded. Peroxide vulcanisation gives optimum heat ageing and compression-set resistance. Unlike organic elastomers, modulus, tensile strength and elongation remain essentially constant over the range 100–200°C. Vulcanisate properties have been described.[123]

Expected applications for PNF include O-rings, seals and vibration dampers where low-temperature flexibility and fuel resistance are important. Good physical strength of PNF over a wide temperature range is an important plus over standard FVMQ, but relatively poor long-term strength retention above 175°C may limit its use in dynamic applications.

14. NITROSO AND CARBOXY NITROSO RUBBERS

A class of fluorinated alternating copolymers of general structure as shown are termed 'nitroso' rubbers.[124]

$$\left[-N-O(C)_x-\right]_n$$

Nitroso rubbers themselves are not readily crosslinkable. Of the many active termonomers investigated, the most successful is a nitroso acid $HO_2C(CF_2)_xNO$ ($x = 2, 3$). Pendent carboxy groups from a few termonomer units permit crosslinking.

Carboxy nitroso rubber (CNR) is such a product. Development quantities of a liquid elastomer have become available.[125]

A typical terpolymer composition is:

$$\left[\begin{matrix} -N-O-CF_2-CF_2- \\ | \\ CF_3 \end{matrix}\right]_{99} \left[\begin{matrix} -N-O-CF_2-CF_2- \\ | \\ (CF_3)_3 \\ | \\ COOH \end{matrix}\right]_{1}$$

Copolymer segment   Crosslink site segment

CNR can be crosslinked by metal oxides and amines, but organometallics such as chromium trifluroacetate provide the best balance of mechanical properties and chemical resistance. Silica-filled vulcanisate tensile strength is in the range 7–14 MPa, and good compression-set resistance can be achieved by an oven post-cure, typically 24 h at 175°C.

SPECIAL-PURPOSE ELASTOMERS

Processing, properties and potential applications of CNR have been reviewed.[126]

Carboxy nitroso rubbers retain many desirable properties of nitroso elastomers, such as low-temperature flexibility down to $-40°C$ and non-flammability, but there is susceptibility of the crosslinks to some solvents and aqueous alkalis, and thermal stability is limited possibly to 190°C for continuous service.

CNR is of interest in aerospace systems where nonflammability is mandatory. Spraying and dip coating can be used to flame-proof such items as electronic components and astronauts' gloves. Potential in moulded goods includes O-rings, valve seats and expulsion bladders. Although permeable to N_2O_4 rocket-fuel oxidiser at 75°C, CNR is resistant to it, unlike FKM or butyl rubber.

15. CARBORANE–SILOXANE POLYMERS

Dating from the early 1960s, these incorporate a carborane ($B_{10}C_2$) nucleus in a polysiloxane chain to enhance thermal stability. A typical SiB-2 polymer comprises a repeating polymer unit of two dimethyl siloxyl groups and one *m*-carobrane moiety. The digit '2' specifies the number of siloxyl groups per carborane unit; SiB-3, 4, etc., have also been synthesised.

$$\left[\begin{array}{c} CH_3 \\ | \\ -Si-CB_{10}H_{10}C- \\ | \\ CH_3 \end{array} \begin{array}{cc} CH_3 & CH_3 \\ | & | \\ Si-O-Si-O- \\ | & | \\ CH_3 & CH_3 \end{array} \right]_n$$

SiB-2 polymer

Modified SiB polymers can be peroxide-cured. Press-cure times depend on the peroxide type. A post-cure of 24 h at 150–250°C is normally required.

Mechanical properties are considerably poorer than for most elastomers, even when silica-reinforced (tensile strength 1·4–4·6 MPa). Metal oxide antioxidants (Fe_2O_3) help prevent surface degradation during post-cure.

Modified SiB polymers can retain useful elastomeric properties after 12 days in air at 315°C.[127] Anticipated service temperatures are as high as 425°C. SiB-2 vulcanisates have high swell in aromatic and chlorinated solvents, and very low swell in alcohols.[128]

Possible applications for these elastomers in the aerospace industry include sealants, gaskets, O-rings and wire and cable insulation. Electrical properties, particularly the low dielectric constant of specially filled vulcanisates, make them attractive insulants. Fluid resistance may need to be improved for use as fuel-tank sealants and fluid-system seals. Low-temperature flexibility of SiB-2 vulcanisates is likely to be satisfactory.

Development quantities of SiB polymers are available in the USA as Dexsil®.

16. NORBORNENE ELASTOMER (NORSOREX®)

16.1. Chemistry

Announced in 1975[129] by Société Chemique des Charbonnages (CdF Chemie) this elastomer is prepared by breaking the norbornene ring to produce a *cis–trans* configuration having the necessary unsaturation to allow crosslinking to take place.

The following reaction was proposed:[129]

(i) cyclopentadiene + ethylene = ⟶ norbornene

(ii)

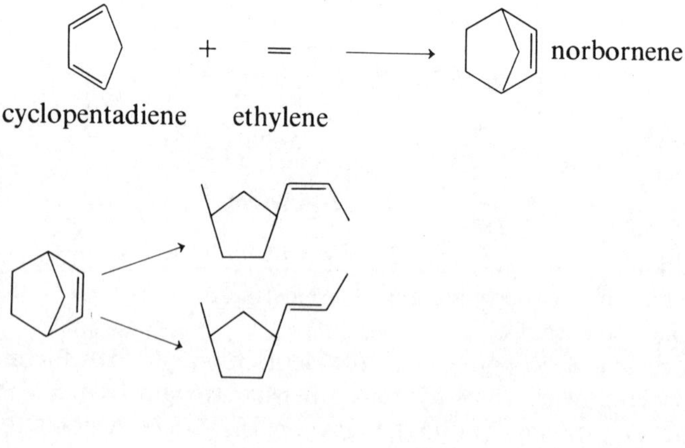

® Dexsil is a registered trade name of Oliss Research Centre, Chemicals Div., Connecticut, USA.

Norsorex is a registered trade name of CdF Chemie, France

16.2. Availability

The polymer in its original form is produced as a fine powder having essentially thermoplastic properties and a second-order transition temperature of +35°C. Mooney plasticity at 100°C is 150, ML 4. However, a very high compatibility with aromatic and naphthenic oils allows compounds to be produced commercially in a wide range of plasticities down to approximately 40, ML 4, with a transition temperature of −60°C. A 5,000-tonne-capacity plant was announced in 1975.

16.3. Compounding

These polymers are notable for their ability to accept very high levels of filler, oil and plasticiser. Vulcanisation can be carried out with conventional rubber-industry curing agents and accelerators although complex EV systems comprising TMTD, TeDEDC, DOTG, ETU, DEDPTDS (diethyl diphenyl thiuram disulphide) and dithiomorpholine in combination are recommended.

16.4. Properties

The norbornene polymers have a useful range of properties including heat resistance to 100°C, ozone resistance if properly protected and reasonable low-temperature properties.

Of particular note is the very high tensile strength obtainable, which is maintained even at a low hardness. For example, 10 MPa at a hardness of 18 Shore A is claimed.[130]

Further, these polymers have good dampening properties over a wide temperature range showing almost constant results between 10°C and 100°C.

16.5. Applications

These will include vibration dampening and highly filled sound-deadening parts for the auto industry. In addition, special low-hardness compounds for printing rolls and stereos can be envisaged.

REFERENCES

1. BROWN, J. H. In *Progress of Rubber Technology*, Vol. 39, PRI, London 1976.
2. NERSASIAN, A. and ANDERSON, D. *J. Appl. Polymer Sci.*, 4 1960, p. 74.
3. TRETYAKOVA, N. M. and KOSMODEMYANSKI, L. V. *Vysokomolekul. Soedin.*, **B-16** (6), 1974, pp. 438–9.

4. *HYPALON® Bulletin No. 2A*, E. I. du Pont de Nemours Inc., January 1969.
5. MAYNARD, J. and JOHNSON, P. *Rubber Chem. Technol.*, **36** (1), 1963, pp. 963–74.
6. DUPUIS, I. C., *Report SD 217*, E. I. du Pont de Nemours, Inc., December 1974.
7. HANDS, B. *Rubber World*, **166** (2), 1972, p. 34.
8. DUPUIS, I. C. *SAE Soc. Automotive Engrs Meeting, Detroit, USA*, May 1973.
9. FREEMAN, D. J. *Paper No. 10, 4th Intern. Symp. Elect. Ins. Mats. Systems, Wroclaw, Poland*, May 1974.
10. LOCKARDT, H. G. GB 1 362 720, publ. August 1974.
11. *One-ply Roofing based on HYPALON®*, E. I. du Pont de Nemours Inc., Ref. A-84008.
12. UK Agrément Board Certificate 74/223.
13. Anon. *Plasty Kaucuk*, **11** (7), 1974, pp. 203–11.
14. RUEBENSAAL, C. F. *Proc. 16th Meeting Intern. Inst. Synthetic Rubber Producers*, May 1975.
15. Dow Chemical Co. US 3 454 544, publ. 8 July 1969.
16. Dow Chemical Co. US 3 429 865, publ. 25 February 1969.
17. Dow Chemical Co. US 3 563 974, publ. 16 February 1971.
18. Osaka Soda Co. Ltd., US 3 759 888, publ. 18 September 1973.
19. *Dow CPE Elastomers*, Dow Chemical Co. Bulletin, 1974.
20. Anon. *Plastics Rubber Weekly*, 7 November 1975, p. 9.
21. ERA, V. *Makromol. Chem.*, **175** (7), 1974, pp. 2191–8.
22. ERA, V. *Makromol. Chem.*, **175** (7), 1974, pp. 2199–202.
23. NAOKI, M. and NOSE, T. *Polymer J. (Jap.)*, **6** (1), 1974, pp. 45–50.
24. NAOKI, M., NAKAJIMA, K., NOSE, T. and HATA, T. *Polymer J. (Jap.)*, **6** (4), 1974, pp. 283–94.
25. HUMBERT, C., BERTICAT, P., QUENUM, B. M. and VALLET, G. *Makromol. Chem.*, **175** (5), 1974, pp. 1611–25.
26. KREUTSEL, L. B. and LITMANOVICH, A. D. *Vysokomolekul. Soedin.*, **169** (5), 1974, pp. 372–4.
27. SAITO, T. and YAMAGUCHI, K. *Polymer*, **15** (4), 1974, pp. 219–27.
28. RHODE, E., *Swedish Rubber Meeting*, October 1978.
29. BLANCHARD, R. R. *J. Elastomers Plastics*, **6**, January 1974, pp. 3–15.
30. NORMAN, R. J. and JOHNSON, J. B. *Paper No. 40, ACS Div. Rubber Chem. 108th Meeting*, October 1975.
31. ABU-ISA, I. A. *SPE 32nd Ann. Tech. Conf., San Francisco*, May 1974.
32. SOLLBERGER, L. E. and CARPENTER, C. B. *Paper No. 9, ACS Div. Rubber Chem. 105th Meeting*, May 1974.
33. BINDER, C. R. and TREXLER, H. E. *Rubber World*, **162** (1), 1970, p. 78.
34. General Electric Co. US 3 738 866, publ. 12 June 1973.
35. Okonite Corp. US 3 821 139, publ. 28 June 1974.
36. Uniroyal, Inc. US 3 772 408, publ. November 1973.
37. IGNATOVA, G. F. et al. *Plast. Massy* (7), 1974, pp. 39–41.
38. Badische Anilin & Soda Fabrik AG. GB 1 351 145, publ. 24 April 1974.
39. SAHAJPAL, V. K. *SPE 31st Ann. Tech. Conf., Montreal*, May 1973.
40. RANKIN, G. M. et al. *Proizv. Shin RTI i ATI* (3), 1974, pp. 4–6.

41. VANDENBERG, E. J. *Rubber Plastics Age*, **46**, 1965, pp. 1139–43.
42. VANDENBERG, E. J. *Paper No. 32, ACS Div. Rubber Chem. 88th Meeting*, October 1965.
43. WILLIS, W. D., AMBERG, L. D., ROBINSON, A. E. and VANDENBERG, E. J. *Rubber World*, **153** (1), 1965, p. 88.
44. HSIEH, H. L. and WEIGHT, R. F. *J. Appl. Polymer Sci.*, **15**, 1971, pp. 2417–24.
45. GURGILIO, A. E. *Rubber Chem. Technol.*, **42** (4), 1969, pp. 1028–33.
46. VANDENBERG, E. J. *J. Polymer Sci.*, **7**, Part A-1, pp. 525–67.
47. B. F. Goodrich Co. GB 1 388 529, publ. 25 March 1975.
48. PAVLYUCHENKO, V. N., BALAEV, G. A. and SOKELOV, V. N. *Plast. Massy* (1), 1974, pp. 5–6.
49. SADYKH-ZADE, S. I., KHANMAMEDOV, T. K. and GASANOV, F. D. *Vysokomolekul. Soedin.*, **B-16** (6), 1974, pp. 465–6.
50. Bayer AG., GB 1 343 823, publ. 16 January 1974.
51. Idemitsu Kosan Co. Ltd. GB 1 342 260, publ. 3 January 1974.
52. TER MEULEN, B. H. *Proc. Intern. Rubber Conf. Prague*, 1973.
53. PFISTERER, H. A. and DUNN, J. R. *Paper No. 40, ACS Div. Rubber Chem. 106th Meeting*, October 1974.
54. COLLINS, E. A. and OETZEL, J. T. *Rubber Chem. Technol.*, **42** (3), 1969, pp. 790–9.
55. COLLINS, E. A. and OETZEL, J. T. *Paper No. 28, ACS Div. Rubber Chem. 95th Meeting*, April 1969.
56. OETZEL, J. T. and COLLINS, E. A. *Paper No. 29, ACS Div. Rubber Chem. 97th Meeting*, April 1970.
57. VANDENBERG, E. J., RALSTON, R. H. and KOCHER, B. J. *Rubber Age*, **102**, 1970, pp. 47–58.
58. VANDENBERG, E. J., RALSTON, R. H. and KOCHER, B. J. *Proc. 4th Intern. Rubber Symp. London*, Sept/Oct 1969.
59. YAMADA, M., ARAI, S. and MASUDA, Y. *Nippon Gomu Kyokaishi*, **46** (5), 1973, pp. 404–10.
60. NAKAMURA, Y., OKA, S., MORI, K. and TAMURA, K. *Nippon Gomu Kyokaishi*, **46** (6), 1973, pp. 507–13.
61. NAKAMURA, Y., MORI, K. and OKA, S., *Nippon Gomu Kyokaishi*, **46** (6), 1973, pp. 514–19.
62. NAKAMURA, Y., OKA, S. and MORI, K. *Nippon Gomu Kyokaishi*, **47** (1), 1974, pp. 48–55.
63. NAKAMURA, Y. *et al.*, *Nippon Gomu Kyokaishi*, **47** (2), 1974, pp. 113–15.
64. Technical brochure, Polysar UK Ltd, Guildford, Surrey.
65. Minnesota Mining & Manufacturing Co., PRAGER, J. H. and MCCURDY, R. M. US 3 031 436, publ. 24 April 1962.
66. MENDELSOHN, M. A. *Ind. Eng. Chem. Prod. Res. Dev.*, **3** (1), 1964, p. 67.
67. Poltmer Corp. Ltd., CHALMERS, D. C. US 3 493 548, publ. 3 February 1970.
68. KING, W. H. *Mach. Des.*, **45** (2), 1973, pp. 106–12.
69. Thiokol Chem. Corp., BERENBAUM, M. B. and KNAVEL, G. A. US 3 335 117, publ. 8 August 1967.
70. Thiokol Chem. Corp., KNAVEL, G. A. and BULBENKO, G. F. US 3 338 876, publ. 29 August 1967.
71. HOLLY, H. W., MIHAL, F. F. and STARER, I. *Rubber Age*, **96**, 1965, p. 565.

72. American Cyanamid Co., MIHAL, F. F. US 3 338 876, publ. 29 August 1967.
73. VIAL, T. M. *Rubber Chem. Technol.*, **44** (2), 1971, pp. 344–61.
74. WEIR, R. J., Polysar Ltd. Paper to invited audience, Mexico City, March 1974.
75. HAGMAN, J. F. and WITSIEPE, W. K. *et al. Paper No. 10, ACS Div. Rubber Chem. 108th Meeting*, October 1975.
76. CARR, J. and GINN, A., *SAE Paper 780403*, March 1978.
77. HAGMAN, J. F. *Data Sheet EA 310.1*, Du Pont.
78. HAGMAN, J. F. *Data Sheet EA 310.2*, Du Pont.
79. POLMANTEER, K. E. *J. Elastoplastics*, **2**, 1970, pp. 165–94.
80. Dow Corning Corp., BRUNER, L. B. US 3 077 456, publ. 12 February 1963.
81. Wacker-Chemie, NITZSCHE, S. and WICK, M. US 3 032 528, publ. 1 May 1962.
82. PRZYBYIA, R. L. *Paper No. 9, ACS Div. Rubber Chem. 104th Meeting*, October 1973.
83. MAKHLIS, F. A. *et al., Kauch. i Rezina* (1), 1975, pp. 23–6.
84. Dow Corning Corp., BRUNER, L. B. US 3 035 016, publ. 15 May 1962.
85. VINCENT, H. L. *Monographs Plastics*, **1**, Part 1, 1972, pp. 385–96.
86. SOUTHWART, D. W. *Paper E-1, Brit. Hydromechanics Res. Assoc. 7th Intern. Conf. Fluid Sealing*, September 1975.
87. PIERCE, D. R. and KIM, Y. K. *Rubber Chem. Technol.*, **44**, (5), 1971, p. 1350.
88. PIERCE, O. R. and KIM, Y. K. *J. Elastoplastics*, **3**, 1971, p. 82.
89. PIERCE, O. R. and YUNG, K. K. *Appl. Polymer Symp.*, **22**, 1973, pp. 103–25.
90. ASTON, M. W. *IRI Symp. Testing*, February 1974.
91. E. I. du Pont de Nemours and Co., PLUNKETT, R. J., US 2 230 654, publ. 1941.
92. DIXON, S., REXFORD, D. R. and RUGG, J. S. *Ind. Eng. Chem.*, **49**, 1957, p. 1687.
93. RUGG, J. S. and STEVENSON, A. C. *Rubber Age*, **82**, 1957, p. 102.
94. E. I. du Pont de Nemours and Co., REXFORD, D. R. US 3 051 677, publ. 1962.
95. E. I. du Pont de Nemours and Co., PAILTHORP, J. R. and SCHROEDER, H. E. US 2 968 649, publ. 1961.
96. Montecatini-Edison SpA, SIANESI, D., BERNARDI, G. C. and REGIO, A. US 3 331 825, publ. 1967.
97. ALEXANDER, J. and OMURA, H. *ACS Div. Rubber Chem. 110th Meeting*, October 1976.
98. PELOSI, L. and HACKETT, E. *ACS Div. Rubber Chem. 110th Meeting*, October 1976.
99. ALEXANDER, J. *Bulletin VT 250 GH. An Extrusion/Atmospheric Pressure Vulcanisation Grade Polymer Offering Very Good Resistance to Steam and Acids*, Du Pont.
100. MORAN, A. L. and PATTISON, D. B. *Rubber World*, **103** (7), 1971, pp. 37–44.
101. GERI, S., GUINCHI, G. and CECCATO, G. *Mat. Plast. Elast.*, **39** (11), 1973, pp. 875–9.
102. MORAN, A. L. *Bulletin VT-310.1. Curative Masterbatches*, Du Pont.
103. GRIFFIS, C. B. and MONTERMOSO, J. C. *Rubber Age*, **77**, 1955, p. 559.
104. ARNOLD, R. G., BARNEY, A. L. and THOMPSON, D. C. *Rubber Chem. Technol.*, **46** (3), 1973, pp. 619–52.

105. THOMPSON, D. C. and BARNEY, A. L. In *Encyclopedia of Polymer Science and Technology*, John Wiley & Sons, 1971.
106. BROWN, D. W. and WALL, L. A. *J. Polymer Sci. Chem. Ed.*, **10** (10), 1972, pp. 2967–82.
107. BARNEY, A. L., KELLER, W. J. and VAN GULICK, N. M. *J. Polymer Sci.*, **8**, 1970, p. 1091.
108. HALLENBECK, A. and MACLACHLAN, J. D. *Bulletin VT 250 GLT. A Low Temperature Fluoroelastomer*, Du Pont.
109. HALLENBECK, A. *Bulletin VT 220 B.70. An Intermediate Low Temperature Fluoroelastomer*, Du Pont.
110. ASTON, M. *PRI Symposium*, Birmingham Univ., December 1972.
111. DWORKIN, D. *Rubber World*, **171** (5), 1975, pp. 43–6.
112. KNOX, R. E. and NERSASIAN, A. *SAE Paper 770867*, September 1977.
113. BLOW, C. M. and LAWRIE, J. M. In *Progress of Rubber Technology*, vol. 37, PRI, London, 1973/4, pp. 41–64.
114. KALB, G. H., QUARLES, R. W. and GRAFF, R. S. *Appl. Polymer Symp.*, **22**, 1973, pp. 127–42.
115. KOCH, S. (Ed.), *Manual for the Rubber Industry*, Farbenfabriken Bayer, July 1970.
116. MITRA, B. C. and KATTI, M. R. *Popular Plastics*, October 1973, pp. 15–22.
117. BLOW, C. M. *Aircraft Eng.*, July 1964, pp. 1–5.
118. KUCKRO, G. W., GREENBERG, D. H., NEWBERG, R. G. and HILL, L. A. *Paper No. 44, ACS Div. Rubber Chem. 104th Meeting*, October 1973.
119. BARTL, H. *Kautsch. Gummi Kunstst*, **25** (10), 1972, pp. 452–5.
120. GAYLORD, N. G. (Ed.). *High Polymers*, Vol. XIII, Part III, Interscience Publishers, 1962.
121. ROSE, S. H. *Polymer Lett.*, **6** (8), 1968, p. 37.
122. TOUCHET, P. and GATZA, P. E. *Paper No. 12, ACS Div. Rubber Chem. 108th Meeting*, October 1975.
123. TATE, D. P. *Rubber World*, **172** (6), 1975, pp. 41–3.
124. HENRY, M. C., GRIFFIS, C. B. and STUMP, E. C. *Fluorine Chem. Rev.*, **1**, 1967.
125. Thiokol Chem. Corp. US 3 637 814, publ. 25 January 1972.
126. LEVINE, N. B. *Rubber Chem. Technol.*, **44** (1), 1971, p. 40.
127. SCHROEDER, H. *et al. Rubber Chem. Technol.*, **39** (4), Part 2, 1966, p. 1186.
128. PETERS, *et al. Paper No. 3, ACS Div. Rubber Chem. 106th Meeting*, October 1974.
129. LE DELLIOU, P. *Sveriges Gummi Tekniska Föreninges 25th Anniversary Meeting*, May 1975.
130. LE DELLIOU, P. *Norsorex and Very Low Hardness*, CdF Chemie, April 1977.

Chapter 3

VULCANISATION SYSTEMS

E. R. Rodger

Monsanto Technical Center, Louvain-La-Neuve, Belgium

SUMMARY

The type of vulcanisation system used to cure the general-purpose and speciality rubbers largely determines the curing and performance characteristics of the chosen rubber. The most significant components varied in the vulcanisation system are the type and level of organic accelerator and level of elemental sulphur. The factors to be considered in the choice of sulphur/accelerator combinations and their influence on processing and curing characteristics and performance of the resultant vulcanisate are discussed.

The structural changes and resultant performance benefits obtained by reduced elemental sulphur are detailed in terms of the semi-EV and EV concept. These systems give considerable benefits in improved oxidative-ageing resistance in natural and synthetic rubbers.

CHEMICAL ABBREVIATIONS

Accelerators
MBT   2-mercaptobenzothiazole
MBTS   dibenzothiazole disulphide
ZMBT   Zinc mercaptobenzothiazole
TMTD   Tetramethylthiuram disulphide
TMTM   Tetramethylthiuram monosulphide
TETD   Tetraethylthiuram disulphide
DPTT   Dipentamethylenethiuram tetrasulphide

DPG	Diphenyl guanidine
DOTG	Di-*o*-tolyl guanidine
ZDMC	Zinc dimethyldithiocarbamate
ZDC	Zinc diethyldithiocarbamate
ZDBC	Zinc dibutyldithiocarbamate
ETU	Ethylene thiourea
TMTU	Tetramethylthiourea
DETU	Diethyl thiourea
DBTU	Dibutyl thiourea
ZBDP	Zinc dibutyl dithiophosphate
ETPT	Bis(diethylthiophosphoryl) trisulphide
BDITD	Bis(diisopropylthiophosphoryl) disulphide
CBS	N-cyclohexylbenzothiazole-2-sulphenamide
TBBS	N-t-butylbenzothiazole-2-sulphenamide
MBS	N-morpholinothio-benzothiazole
DCBS	N,N dicyclohexyl benzothiazole-2-sulphenamide
OTOS	N-oxydiethylenethiocarbamyl-N-oxydiethylene sulphenamide
DTDM	4,4′ dithiodimorpholine
ETPT	Bis(diethyl thiophosphoryl) trisulphide
DTBC	N,N′-dithiobishexahydro-2,4-azepinone
BDTM	2-benzothiazole dithio-N-morpholine

Antidegradants
6PPD	N-1,3dimethylbutyl-N′phenyl-*p*-phenylenediamine
IPPD	N-isopropyl-N′-phenyl-*p*-phenylenediamine
TMQ	polymerised 2,2,4-trimethyl-1,2-dihydroquinoline

Retarders
CTP	N-cyclohexylthiophthalimide
NDPA	N-nitrosodiphenylamine

1. INTRODUCTION

The vulcanisation system is the collection of additives required to vulcanise an elastomer, to transform essentially linear polymer molecules into a three-dimensional network by the insertion of crosslinks. The basic objective in the development of a vulcanisation system is the production of crosslinks with required physical and chemical properties. Although most vulcanisation systems consist of only 0·5–5% by weight of the compound, they play a major role in the attainment of required standards of performance by the most economic means of production. The majority of systems in use involve the generation of sulphur-containing crosslinks, usually with elemental sulphur in combination with an organic accelerator. In recent years, the proportion of sulphur has tended to fall and the levels of accelerator and the use of sulphur donors increased to give great improvements in the thermal and oxidative stability of the vulcanisate. Other vulcanisation systems not involving sulphur or sulphur donors are less commonly used and involve urethanes, peroxides, metal oxides and resins.

In recent years the rate of introduction of new chemicals for vulcanisation has tended to slow down, due in large part to increasingly strict environmental and health standards and the inevitable influence on development time scales and economics. This has tended to focus attention in the rubber industry on the efficient utilisation of existing products and exploiting the benefits of the most effective combinations.

2. SULPHUR VULCANISATION

Sulphur was the first agent used to vulcanise the first commercial elastomer, natural rubber (NR). Vulcanisation was achieved by mixing eight parts of sulphur per hundred parts (phr) of polymer and required 5 h at 140°C. The addition of metallic oxides (zinc oxide 5 phr), fatty acid and finally organic accelerators (0·5–2 phr) completed the make-up of the modern vulcanisation system, giving cure times as low as 2–5 min. The consequent reduction in requirement for free sulphur (2–3 phr) produced substantial improvements in physical and performance characteristics.

Accelerated sulphur vulcanisation is suitable for the following types of elastomers:

1. *General purpose*
   (a) Natural rubber (NR)
   (b) Synthetic isoprene rubber (IR)
   (c) Polybutadiene rubber (BR)
   (d) Styrene/butadiene rubber (SBR)
2. *Speciality*
   (a) Nitrile rubber (NBR)
   (b) Butyl rubber (IIR)

(c) Chloro-butyl rubber (CIIR)
(d) Bromo-butyl rubber (BIIR)
(e) Ethylene/propylene/diene modified rubber (EPDM)

The basic recipe for the vulcanisation system is

Zinc oxide	2–10 phr†
Stearic acid	1–4
Sulphur	0·5–4
Accelerator	0·5–10

Zinc oxide and stearic acid comprise the common activator system where the zinc ions are made soluble by salt formation between the acid and the oxide. The part of the vulcanisation system that offers the most opportunity for variation is the sulphur level and type and level of organic accelerator.

2.1. Mechanism of Sulphur Vulcanisation

Accelerated sulphur vulcanisation is thought to proceed by the following steps:[1]

1. The accelerator reacts with sulphur to give monomeric polysulphides of the type $Ac-S_x-Ac$, where Ac is an organic fragment derived from the accelerator. Certain initiating species may be necessary to start the reaction, which then appears to be autocatalytic.
2. The polysulphides can interact with rubber to give polymeric polysulphides of the type rubber$-S_x-$Ac. During this reaction, the formation of mercaptobenzothiazole (MBT) was observed when an accelerator derived from MBT had been used. When MBT itself is used, it first disappears, then appears during the formation of rubber polysulphides.[2]
3. The rubber polysulphides then react, either directly or through a reactive intermediate, to give crosslinks or rubber polysulphides of the type rubber$-S_x-$rubber. If a sulphenamide accelerator is used, the reaction can be represented as in Fig. 1.

† Concentrations given in parts per 100 parts of elastomer.

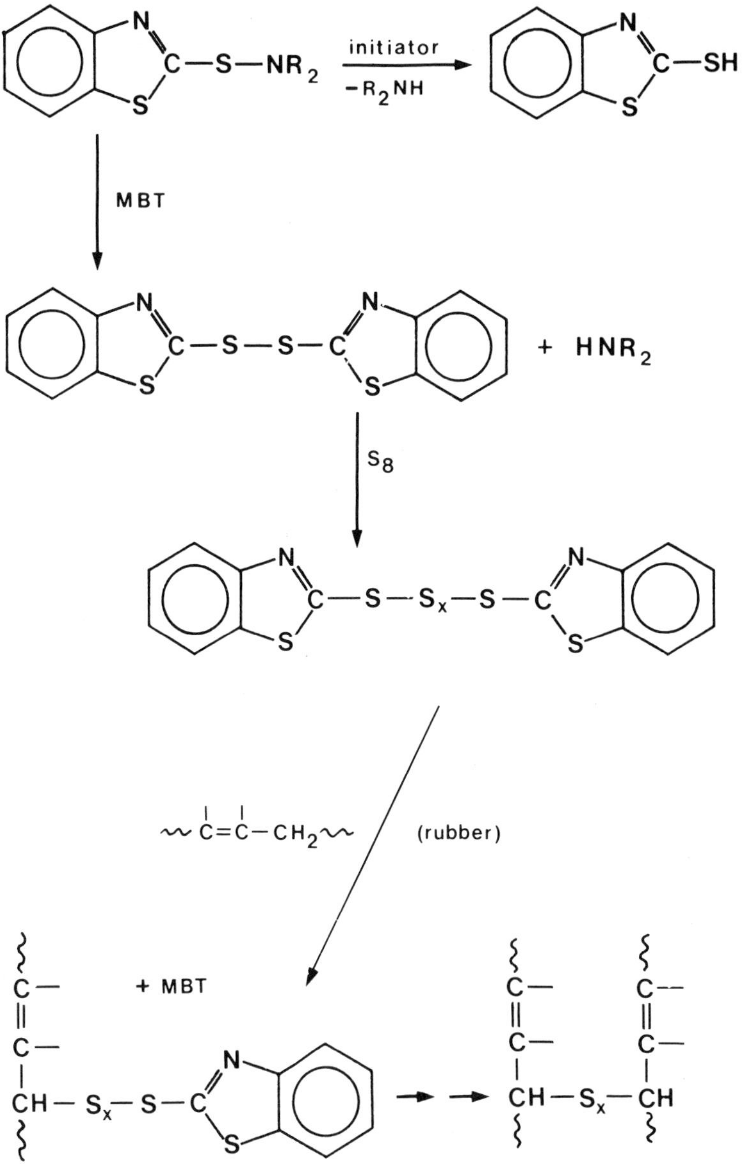

FIG. 1. Sulphur/sulphenamide vulcanisation.

Crosslinking in the presence of a sulphenamide accelerator does not start until virtually all the sulphenamide has reacted.[3] In the absence of accelerators, large amounts of sulphur are required to achieve a given state of cure and the rate of crosslinking is extremely slow. However, more subtle differences exist and work with model olefins suggests that in the case of accelerated sulphur vulcanisation, the sulphur attacks the rubber hydrocarbon almost exclusively at the allylic positions.[4] If no accelerator is used, most of the substitution is on other carbon atoms.[5] These observations suggest that the effectiveness of accelerators lies in the mobilisation of the allylic positions.

2.2. Parameters of Vulcanisation

The critical parameters in vulcanisation are the time before onset and the rate and extent to which it occurs. There must be sufficient delay (scorch time or processing safety) before the onset of vulcanisation to permit mixing, forming and moulding. Once vulcanisation starts, it should be rapid and its extent controlled.

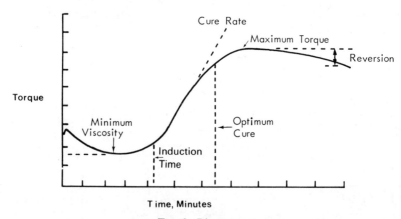

FIG. 2. Rheometer curve.

Scorch time is usually measured by the time at any given temperature required for the onset of crosslink formation as measured by a rapid increase in viscosity. The most common instrument used is the Mooney viscometer. This consists of a continuously rotating disc in a heated cavity which contains the test sample. The temperature is selected to be a typical or average processing temperature (commonly 121°C or 135°C). Extent and rate of vulcanisation are usually measured in cure meters which follow the resistance to motion of an oscillating rotor embedded in the test sample, held at a chosen temperature (140–190°C). These instruments follow cure from the initial phase through crosslinking to the fully cured stage.[6] The resistance to oscillation through a small degree of arc (1° or 3°) is measured and recorded as a function of time to give a characteristic cure curve for the individual test sample (Fig. 2).

Although these instruments measure a viscoelastic rather than a failure property of the vulcanisate, the final cured torque value can be related to modulus measured by tensile testing methods for individual formulations.[7]

2.3. Effects of Vulcanisation on Vulcanisate Properties

Major effects of vulcanisation[8–10] are illustrated by the representation in Fig. 3. It should be noted that the static modulus increases with vulcanisation to a greater extent than the dynamic modulus.[11] The dynamic modulus is a composite of viscous and elastic responses, whereas the static modulus is a measure of the elastic component alone. Vulcanisation thus causes a shift from viscous or plastic behaviour to elasticity.

Tear strength, fatigue life and toughness are related to the energy at break. These properties increase with small amounts of crosslinks but are reduced with increasing crosslink formation. Hysteresis diminishes with increasing crosslink formation and is a measure of the deformation energy that is not stored or borne by the network chains, but instead is converted to heat. Properties related to energy at break then increase with increases in the number of network chains and hysteresis; but since hysteresis decreases as more network chains are developed, the energy-to-break-related properties peak at some intermediate crosslink density.

The properties of Fig. 3 are not functions of crosslink density alone but are also affected by the type of crosslink, the nature of the polymer, the type and amount of filler, etc.

Reversion is a term applied to the loss of modulus by non-oxidative thermal ageing. It is usually associated with isoprene rubbers vulcanised by sulphur and can be the result of long vulcanisation or of hot ageing of thick sections. It is most severe at temperatures above 150°C in vulcanisates containing a large number of polysulphidic crosslinks (see Section 3).

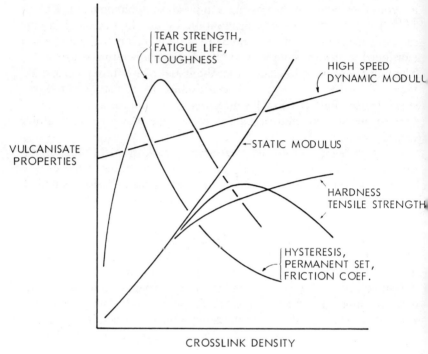

Fig. 3. The effects of vulcanisation.

3. CROSSLINK STRUCTURE

Once the basic polmer has been selected, performance characteristics of the vulcanisate are determined principally by the amount and type of crosslinks produced by the vulcanisation system. A detailed knowledge of the principal features of the crosslink structure and the factors which influence their generation is of practical relevance in designing vulcanisation systems to meet performance and production criteria.

Direct analysis of vulcanisates is extremely difficult due to their insolubility and other reasons. Nevertheless using a variety of approaches, it has been possible to define in some detail the structure of vulcanised diene elastomers and in particular natural rubber. The use of chemical probes has enabled deductions to be made on the types and extent of crosslinks present, although they give only limited information on the position of the crosslink junction point with the polymer chain. In general these methods involve the determination of the number of combined sulphur atoms per crosslink before and after treatment with probes which eliminate certain structures. Before treating the vulcanisate with a chemical probe, some initial information on the fate of the sulphur combined during vulcanisation may be obtained by calculating the Moore–Trego efficiency parameter E given by

$$E = \frac{\text{gram atoms of total combined sulphur/gram rubber}}{\text{gram molecules of chemical crosslinks/gram rubber}}$$

Some typical results[12] are given in Table 1. These values indicate the inefficiency of sulphur usage in the unaccelerated sulphur system, where 40–55 sulphur atoms are required for each crosslink formed and contrast with the efficient usage of sulphur in the TMTD-accelerated system without free sulphur. Redetermination of E after treating the network with triphenylphosphine (which reduces all di- and polysulphidic crosslinks to monosulphides) gives a second parameter E'. The value $E' - 1$ now represents the number of sulphur atoms combined in the network not involved in chemical crosslinks, i.e. roughly the number of gram atoms of sulphur involved in main-chain modifications.

TABLE 1

EFFICIENCY PARAMETER E

Vulcanisation system	Cure temperature, °C	Range of E values
Unaccelerated (NR 100, sulphur 6–10 phr)	140	40–55
Sulphenamide-accelerated (NR 100, sulphur 2·5, CBS 0·6, zinc oxide 5, lauric acid 0·7)	140	12–22
Thiuram-accelerated—sulphurless (IR 100, TMTD 4·0, zinc oxide 4, lauric acid 1·5)	140	2·1–7·9

The various structures now known to be present in a sulphur-vulcanised natural rubber network are shown in Fig. 4.

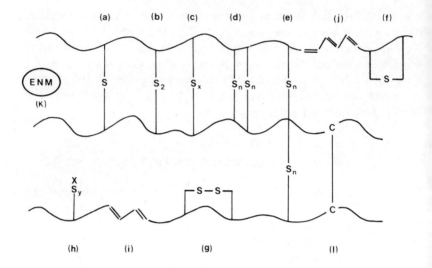

Fig. 4. Network structure. a, Monosulphide crosslink C—S—C; b, disulphide crosslink C—S—S—C; c, polysulphide crosslink C—S_x—C($3 \leq x < 6$); d, parallel vicinal crosslink C—S_n—C(n=1 to 6); e, crosslinks attached to common or adjacent carbon atoms; f, intra-chain cyclic monosulphide; g, intra-chain cyclic disulphide; h, pendent sulphidic group terminated by accelerator fragment (X); i, conjugated diene; j, conjugated triene; k, extra-network material; l, carbon–carbon crosslink (probably absent).

Whereas crosslinks a–e contribute to the strength and performance of the vulcanisate, the wasted sulphur in main-chain modifications (f–h) is believed to contribute to poor oxidative-ageing resistance as do the conjugated dienes and trienes. As indicated in Table 2, the choice of type and level of accelerator and of sulphur level has a large influence on the efficiency of sulphur usage in the network and also the type and distribution of crosslinks. This is illustrated in Table 2 for an efficient system with no free sulphur and high total amount of accelerators and sulphur donor DTDM compared with a system with a conventional level of sulphur and low accelerator level.

The distinctive features of an efficient vulcanisation (EV) system can be clearly seen. The proportion of short mono- and disulphide crosslinks is high and that of polysulphides low. In terms of resistance to thermal ageing (reversion) this is beneficial, as the long polysulphidic crosslinks tend to rearrange after long cure times or at high tempera-

TABLE 2
COMPARISON OF CROSSLINK STRUCTURES OF CONVENTIONALLY AND EFFICIENTLY VULCANISED NR[a]

	Conventional (sulphur 2·5, CBS 0·5)	Efficient (DTDM 1·0, CBS 1·0, TMTD 1·0)
Initial crosslink density $(2M_c)^{-1} \times 10^5$ [b]	5·8	4·13
% Monosulphidic	0	38·5
% Disulphidic	20	51·8
% Polysulphidic	80	9·7
E	10·6	3·5
E'	6·0	3·0
$E' - 1$	5·0	2·0

[a] Formulation: NR 100, zinc oxide 5, stearic acid 1.
[b] $(2M_c)^{-1}$ = moles of chemical crosslinks per gram of rubber network.

tures to shorter crosslinks but with a net loss in crosslink density (and hence modulus). The proportion of sulphur not involved in crosslinks ($E' - 1$) is low and this generally leads to improved oxidative-ageing resistance.

This work initially carried out in gum (unfilled) vulcanisates has since been shown to be applicable to black-filled compounds.[13] Although the majority of the determinations of network structure have been carried out in natural rubber they do apply broadly to the other sulphur-vulcanisable elastomers.

4. SELECTION OF VULCANISATION SYSTEM

As discussed in previous sections, in the components of the vulcanisation system—metal oxide, fatty acid, sulphur and accelerator—the degree of choice in the first three is generally a question of level for zinc oxide, stearic acid and elemental sulphur. For organic accelerators, there is an almost bewildering variety.

4.1. Accelerators

European Rubber Journal published a list of accelerators in production and use in 1973 giving 115 single products of known composition and 38 blends and unspecified materials. They may, however, be classified into seven main groups according to their chemical composition and speed of vulcanisation (Table 3). With these accelerators or their combinations it is possible to achieve almost any required processing safety, cure time and modulus.

TABLE 3

ACCELERATOR GROUPS

	Abbreviation	Speed
Guanidines	DPG, DOTG	slow
Dithiocarbamates	ZDMC, ZDC, ZDBC	very fast
Thiurams	TMTD, TMTM, TETD, DPTT	very fast
Thioureas	ETU, TMTU, DETU, DBTU	fast
Thiophosphates	ZBDP, ETPT, BDITD	semi-fast
Thiazoles	MBT, MBTS, ZMBT	moderate
Sulphenamides	CBS, TBBS, MBS, DCBS	fast

The features required of an ideal accelerator can be summarised as

1. fast cure;
2. high activity (crosslinking efficiency);
3. soluble in rubber (no bloom, good dispersion);
4. delayed action (good processing safety);
5. good storage stability (as 100% material and compounded);
6. flat plateau (no reversion—particularly in NR);
7. effective over a wide range of temperatures;
8. compatible with other ingredients;
9. no safety or handling problems;
10. no side effects on other properties (ageing, adhesion, etc.).

The accelerators which have the widest application worldwide and are produced in the greatest volume are the thiazoles and sulphenamides. Of these, sulphenamides have the greatest range due in large part to the combination of fast cure, good processing safety, high activity and

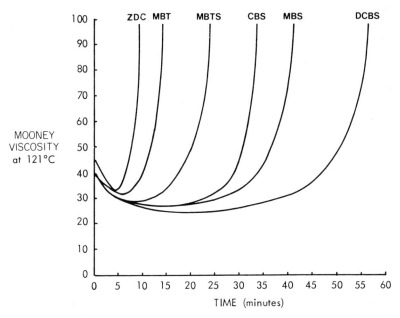

FIG. 5. Effect of cure system on scorch time. Formulation: NR 100, N330 50, processing oil 8, zinc oxide 5, stearic acid 2, 6PPD 2, sulphur 2·5.

solubility. The thiazoles, however, confer slightly greater resistance to thermal and oxidative ageing to the vulcanisate. Basic processing and curing characteristics are compared in Figs. 5 and 6 for three sulphenamides (CBS, MBS, DCBS), a thiazole (MBTS) and dithiocarbamate (ZDC) in a 100 NR formulation.

4.2. Recent Accelerator Developments

There have been relatively few products developed in recent years outside the groups in Table 3. One significant new group has been the triazine accelerators[14] (e.g. bis(2-ethylamine-4-diethylaminotriazine-6-yl)). These products have a higher efficiency than existing accelerators and give good processing safety with fast rates of cure. Lower loadings are possible than used with the sulphenamides. They do, however, normally require activation to be most effective and have not been adopted to any great extent.

Solubility problems associated with thiuram accelerators normally

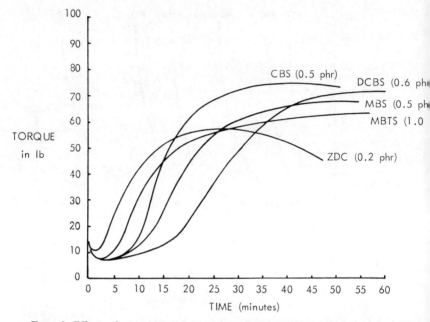

FIG. 6. Effect of cure system on cure characteristics. Rheometer at 140°C. Formulation: NR 100, N330 50, processing oil 8, zinc oxide 5, stearic acid 2, 6PPD 2, sulphur 2·5.

used in EPDM and its blends with NBR and SBR led to the proposed use of long-chain hydrocarbon-substituted accelerators, in particular bisdodecyl- and bisoctadecylisopropylthiuram disulphides.[15]

A recent development[16] in sulphenamide accelerators has been that of the thiocarbamyl sulphenamides such as N-oxydiethylenethiocarbamyl-N-oxydiethylene sulphenamide (OTOS). This has good processing safety, a fast cure and high modulus development but can give NR vulcanisates with inferior oxidative-ageing resistance to CBS or TBBS.

Two additional components of the vulcanisation system which should also be considered are sulphur donors and retarders.

4.3. Sulphur Donors

In Section 3 it was seen that reductions in free sulphur coupled with increasing accelerator level gave vulcanisates with essentially mono- and disulphidic crosslinks. This was coupled with an efficient usage of sulphur in the network and a low degree of main-chain modifications. Progressive reductions in free sulphur require increases in accelerator level if modulus and other properties are to be maintained. When sulphur is reduced to low levels (below 0·5 phr) it is not possible to match modulus simply by increasing accelerator, unless the accelerator is of a type which can donate sulphur for crosslinks. A number of accelerators (TMTD, TETD, DPTT, BDTM, ZBDP) can function both as accelerators and sulphur donors. A number of other products are able to donate sulphur but do not function as accelerators and must always be used in conjunction with an accelerator (DTDM, ETPT, DTBC). Structures for the two most common sulphur donors are shown below.

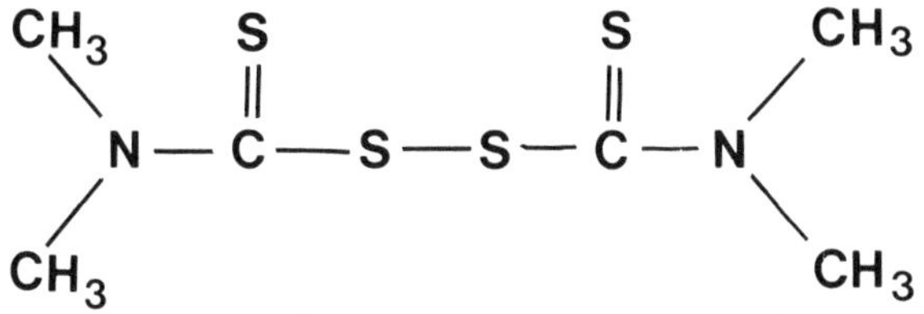

TMTD (sulphur donor/accelerator)

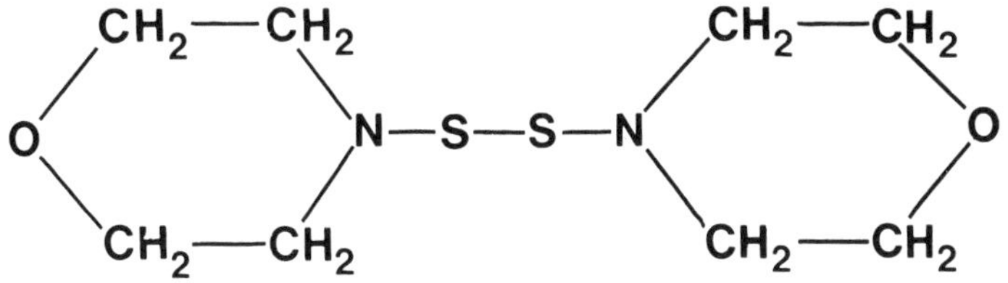

DTDM (sulphur donor)

In systems that contain approximately 0·5 phr of free sulphur or more, increasing the level of accelerator will enable modulus to be maintained with moderate levels of accelerator. Below 0·5 phr sulphur, modulus is less responsive to accelerator level. This is illustrated for NR in Fig. 7.

Generally sulphur donors are used where the free-sulphur level is reduced with the objective of improving thermal and oxidative-ageing resistance. There are, however, cases where it is required to lower the sulphur level in a compound to eliminate the possibility of bloom. Sulphur donors can also be used to modify processing and curing characteristics as seen in Table 4 for DTDM in black-filled NR.

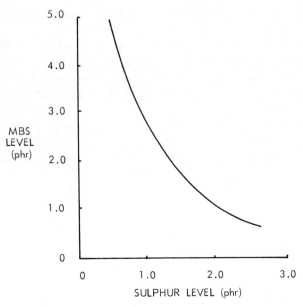

FIG. 7. Sulphur/accelerator levels for equal modulus in 100 NR.

TABLE 4

EFFECT OF DTDM ON PROCESSING AND CURE CHARACTERISTICS IN NR

MBS	0·6	0·6	—	—
MBTS	—	—	0·9	0·9
DTDM	—	0·6	—	0·6
Sulphur	2·5	1·5	2·5	1·5
Processing characteristics, Mooney at 121°C, scorch time t_5 (min)	24	39	14	20
Curing characteristics, rheometer at 153°C Cure time t_{90} (min) Maximum modulus (in lb)	32 64	32 64	24 58	20 58

4.4. Retarders

Retarders have been a useful addition to the vulcanisation system for many years as a short-term solution to scorch problems. They are added to the formulation specifically to obtain longer processing safety t avoid premature cure in factory processing. Two principle types are available:

1. acids—phthalic anhydride, salicylic acid, benzoic acid;
2. nitroso compounds—N-nitrosodiphenylamine (NDPA).

Retarders, where required, are added at 0·5–2·0 phr; however, the acidic types in particular can have side effects in slowing down rate of cure and modifying modulus.

The development of prevulcanisation inhibitors in 1970 solved most of the problems associated with the existing retarders. The most widely used product in this class—N-cyclohexylthiophthalimide (CTP)(Fig. 8)—increases processing safety by an amount directly proportional to the quantity added, with minimal effects on curing or physical properties and performance characteristics. According to Leib et al.,[17] CTP scavenges MBT which is an autocatalyst for the breakdown of the accelerator to form cyclohexyldithiobenzothiazole (CDB) and phthalimide.

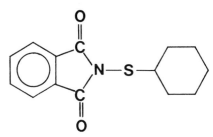

FIG. 8. CTP structure.

The main applications of prevulcanisation inhibitors have been in the following areas:[18,19]

1. Improved mixing efficiency—single, two-stage and continuous.
2. Improved storage stability of mixed compounds.
3. Elimination of short-term processing problems.
4. Replacement of conventional retarders.
5. Processing with scorch-reducing additives.
6. Increased calendering/extrusion rates.
7. Reduction of cure times.
8. Improved bonding to brassed steel wire.
9. Recovery of partly scorched compounds.
10. Improved mould flow with thick sections.

5. COMPOUNDING WITH ACCELERATORS

As pointed out in Section 4.1, the use of existing accelerators alone or in combinations enables almost any required level of processing and curing characteristics to be achieved. Indications of starting point vulcanisation systems for black-filled NR, SBR, NBR, IIR and EPDM are given in Table 5.

TABLE 5

CONVENTIONAL VULCANISATION SYSTEMS

	\multicolumn{5}{c}{Polymer}				
	NR	SBR	NBR	IIR	EPDM
Sulphur	2·5	2·0	1·5	2·0	1·5
Zinc oxide	5·0	5·0	5·0	3·0	5·0
Stearic acid	2·0	2·0	1·0	2·0	1·0
TBBS	0·6	1·0	—	—	—
MBTS	—	—	1·0	0·5	—
MBT	—	—	—	—	1·5
TMTD	—	—	0·1	1·0	0·5

These 'conventional' vulcanisation systems are all based on normal sulphur levels for the respective polymers and are meant as an indicator for comparative purposes only. The individual vulcanisation system for a particular application will be designed to achieve required processing and curing characteristics and is specific to a given polymer/filler system and is affected by changes in other compounding ingredients such as processing oils, antidegradants, etc.

For a given single accelerator, substantial variations in processing and curing characteristics can be achieved by changing the ratio of accelerator to sulphur (Fig. 9, Table 6) with constant modulus. With NR (Fig. 9), simply changing the curing system from 0·6 MBS/2·5 sulphur to 2·0 MBS/1·2 sulphur gives a 15-min increase in scorch time with the advantage of a shorter cure time. However, even when there is a single primary accelerator this approach is not predictable as it can not be applied to all polymers. With SBR,[20] for example, the advantages of the high accelerator/low sulphur system are not as pronounced as with NR although cure time is reduced, processing safety is also reduced (Table 6). The low response of modulus to increasing sulphenamide and the increase in cure time at low sulphur level is particularly marked.

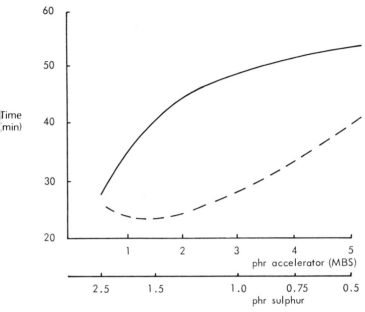

FIG. 9. Effect of accelerator level on scorch and cure characteristics in NR. ——Mooney scorch at 121°C;----t_{90} at 141°C.

Increasing accelerator level at a fixed level of sulphur will also modify curing characteristics (Table 7), as well as increasing modulus. In this case cure time is progressively reduced with practically no change in processing safety.

TABLE 6

INCREASING ACCELERATOR IN SBR[a]

Sulphur	2·0	1·5	1·0	0·75
TBBS	1·0	1·6	3·25	5·25
Processing characteristics, Mooney at 135°C, scorch time t_5 (min)	28	27	23	22
Curing characteristics, rheometer at 153°C				
Maximum modulus (in lb)	82	80	80	81
Cure time t_{90} (min)	29	22·5	19	24

[a] Formulation: SBR-1500 100, N330 50, high aromatic oil 8, zinc oxide 4, stearic acid 2.

TABLE 7

EFFECT OF ACCELERATOR LEVEL ON SCORCH AND CURE CHARACTERISTICS IN NR[a]

Sulphur	2·5	2·5	2·5	2·5
MBS	0·5	0·75	1·0	1·25
Processing characteristics, Mooney at 121°C, scorch time t_{10} (min)	32	32	34	32
Curing characteristics, rheometer at 141°C Cure time t_{90} (min)	34	26	19	18

[a] Formulation: NR 100, HAF 45, processing oil 3·5, zinc oxide 5, stearic acid 2, TMQ 1·0.

Fast-curing accelerators (thiurams, dithiocarbamates, etc.) are commonly used to achieve faster cure times than is possible with single accelerators whilst maintaining a required degree of processing safety.[21-23] An example is shown in Fig. 10.

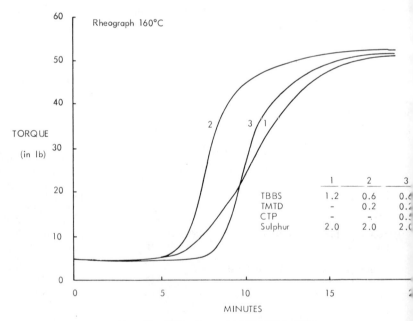

FIG. 10. Activation with TMTD in SBR.

The increase in modulus due to the addition of the fast-curing accelerator can be compensated for by reducing the level of primary accelerator if equal modulus is required. Cure time is reduced (system 2) but usually with a reduction in processing safety. To maintain the same processing safety, the addition of a retarder or prevulcanisation inhibitor (in this case CTP) is required. The resulting activated sulphenamide system with the addition of CTP (system 3) will enable processing safety to be maintained or even increased whilst still achieving a shorter cure time.

A further example of the use of accelerator combinations to achieve optimum processing and curing characteristics is shown in Table 8. An activated sulphenamide system can replace a thiazole system with lower total accelerator and faster cure time while matching physical properties and processing safety.

TABLE 8

COMPARISON OF ACTIVATED SULPHENAMIDE WITH THIAZOLE IN NR/SBR

Sulphur	2·5	2·5
MBT	0·6	—
MBTS	0·9	—
MBS	—	0·6
TMTD	—	0·3
Processing characteristics, Mooney at 135°C, scorch time t_5 (min)	9·0	9·5
Curing characteristics, rheometer at 160°C, Optimum cure time t_{90} (min)	6·1	4·4
Physical properties		
Tensile strength ($kg\,cm^{-2}$)	132	131
300% Modulus ($kg\,cm^{-2}$)	132	130
Elongation at break (%)	300	310

6. EV AND SEMI-EV SYSTEMS

The terms efficient and inefficient have been outlined earlier in terms of the usage of sulphur in the vulcanised network. The terms EV (efficient vulcanisate or efficient vulcanisation) and inefficient, or conventional, represent two extremes. At one extreme, the EV uses a low or even zero level of sulphur, a high level of accelerator and possibly a sulphur donor. In NR the resultant vulcanisate has a high proportion of mono and disulphide crosslinks and a low degree of main-chain modification, giving high resistance to thermal and oxidative ageing. The conventional (for NR) system has a high proportion of main-chain modifications. In NR this gives poor resistance to thermal and oxidative ageing.

The term EV is similarly applicable to other sulphur-vulcanisable polymers. However, except for SBR[24,25] and to some extent NBR,[2] there is little information available on the structure of their vulcanisates. For SBR, structural determinations show that even the conventional system has a high proportion of monosulphide crosslinks (Table 9). The thermally labile polysulphide crosslinks are present to a very low extent in SBR EV systems compared with the more stable monosulphides.

TABLE 9

CROSSLINK DISTRIBUTIONS FOR NR AND SBR

	Crosslink type (%)			
	NR		SBR	
Vulcanisation system	S_1	$S_2 + S_x$	S_1	$S_2 + S_x$
Conventional[a]	0	100	38	62
EV[b]	46	54	86	14

[a] Conventional for NR: sulphur 2·5, MBS 0·6; conventional for SBR: sulphur 2·0, CBS 1·0.
[b] EV for NR: CBS 1·5, DTDM 1·5, TMTD 1·0; EV for SBR: CBS 1·5, DTDM 2·0, TMTD 0·5.

A further difference between NR and SBR and the other synthetic polymers is in the mechanism of oxidative ageing and lack of reversion. In SBR, oxidative ageing causes little main-chain scission (main mechanism of oxidation in NR) but a high degree of additional crosslinking[25,27] resulting in increased modulus and hardness and reduced elongation at break. These oxidatively induced crosslink changes have been followed by evaluations in SBR-filled vulcanisates as in NR. In addition to the sulphur-containing crosslinks found in NR, a further non-sulphidic crosslink has been detected in SBR, referred to in Fig. 11 as —X—, the exact nature of which is unclear but which in

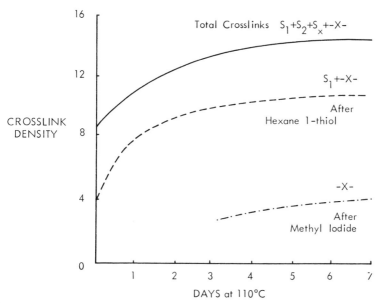

FIG. 11. Crosslink density v. ageing time in conventional SBR. Sulphur 2·0, CBS 2. Crosslink density = $(2M_cPhys)^{-1} \times 10^5$; M_cPhys = number average molecular weight of network chains between effective crosslinks.

creases with oxidative ageing. The proportions of crosslinks of each type were determined by measurements of crosslink density before and after treatment with chemical probes. These probes selectively cleave specific types of crosslinks:

$S_2 + S_x$ (di- and polysulphides)—hexane-1-thiol.

$S_1 + S_2 + S_x$ (mono- di- and polysulphides)—methyl iodide.

Substantial improvements in oxidative-ageing resistance can be obtained with EV systems in synthetic polymers. The behaviour of a sulphurless EV SBR during oxidative ageing[20] is shown in Fig. 12.

The substantial reduction in non-sulphidic crosslinks found in EV-cured SBR is reflected in the reduced modulus increases during ageing (Fig. 13).

A further difference in EV systems between NR and the synthetic polymers is the behaviour of fatigue life[13,28,29] as the level of sulphur in

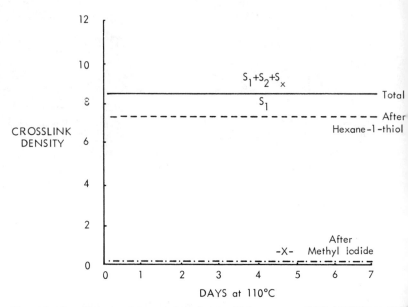

FIG. 12. Crosslink density v. ageing time—efficient system SBR. DTDM 2, CBS 1·5, TMTD 0·5.

the vulcanisation system is reduced (Fig. 14). In the synthetic polymers a reduction is not observed; on the contrary, in many cases, EV systems with reduced or no elemental sulphur give significantly higher fatigue lives than conventional systems.

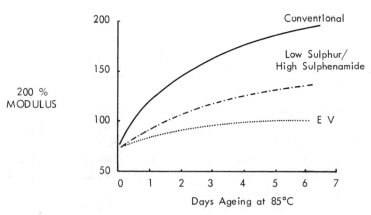

FIG. 13. Marching modulus in SBR.

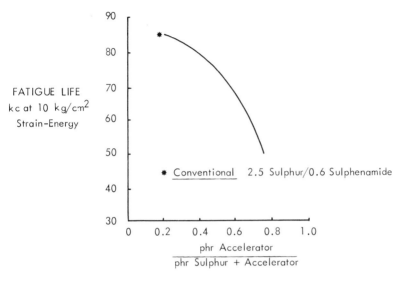

FIG. 14. Fatigue life (measured by extension) in NR—unaged.

Semi-EV systems. The term 'semi-EV' has been applied to vulcanisation systems with sulphur levels intermediate between the conventional and EV. Such semi-EV systems are used widely for compromises in costs and/or performance. There is, however, no unique level of accelerator and sulphur contents making the transition from a conventional system to a semi-EV or from a semi-EV to an EV. The semi-EV system has found particular application in NR where there is a need to compromise between improvements in thermal and oxidative ageing and decreasing fatigue life as sulphur level is reduced. Examples of semi-EV vulcanisation systems are given for NR and SBR based on sulphur/sulphenamide in Table 10 with all systems designed to give equal cured modulus.

6.1. Compounding and Performance of EV and Semi-EV Systems

EV and semi-EV systems are used principally to obtain improvements in oxidative-ageing resistance (and in NR also reversion resistance); however, other benefits are also obtained such as reduced compression set. The greatest resistance to oxidative ageing and lowest compression set is obtained with peroxide-cured vulcanisates (e.g. dicumyl peroxide).

TABLE 10
CURING SYSTEM FOR EQUAL MODULUS

	Conventional	Semi-EV	EV
in NR			
Sulphur	2·5	1·5	0·5
Sulphenamide	0·6	1·5	5·0
in SBR			
Sulphur	2·0	1·2	0·75
Sulphenamide	1·0	2·5	7·0

However, their application is strictly limited due to numerous drawbacks—inflexible processing safety, odour and generally poor initial vulcanisate properties in addition to their high cost and the necessity for careful storage.

A more detailed discussion of the semi-EV and EV systems used in the major sulphur-vulcanisable polymers and the benefits in performance obtained is given in the following sections.

6.1.1. EV and Semi-EV Systems in Natural Rubber

As discussed previously the reduction in fatigue life in NR as elemental sulphur level in the vulcanisation system is reduced led to interest in semi-EV systems, in particular for applications where fatigue resistance is critical (e.g. tyres). Semi-EV systems in NR usually contain approximately 1·5 phr sulphur with either increased level of accelerator or a sulphur donor[13] (Table 11). In this case both routes to the semi-EV system give similar longer processing safety than the conventional with shorter cure times. The use of higher accelerator only, however, tends to give poorer stability on storage. Relative performance on ageing and reversion resistance is shown in Table 12. Both semi-EV systems give comparable improvements in reversion and ageing resistance after overcure.

Exactly the same principles apply to semi- and fully efficient vulcanising systems based on either DTDM or high accelerator where final curing characteristics and ultimate ageing performance is governed to some extent by the choice of primary accelerator (thiazole or sulphenamide). This is illustrated in Table 13. In both conventional and semi-EV systems the use of MBTS gives a shorter scorch time with a concomitant reduction in cure time. Physical properties are given in Table 14. Physi-

TABLE 11

SEMI-EV SYSTEMS IN NR[a]

	Conventional	Sulphur donor	High accelerator
Sulphur	2·5	1·5	1·5
MBS	0·6	0·6	1·5
DTDM	—	0·6	—
Processing characteristics Mooney at 121°C, scorch time t_5 (min)	27	35	32
Curing characteristics rheometer at 141°C			
Optimum cure time t_{90} (min)	25	21	19
Maximum modulus (in lb)	69	67	68

[a] Formulation: NR 100; N330 50; zinc oxide 5; stearic acid 2; processing oil 3; TMQ 1·0; IPPD 1·0.

TABLE 12

EFFECT OF CURING SYSTEM ON OVERCURE IN NR

	Conventional	High accelerator	Sulphur donor
MBS	0·6	1·5	0·6
DTDM	—	—	0·6
Sulphur	2·5	1·5	1·5
Physical properties—optimum cure at 140°C			
Tensile strength ($kg\,cm^{-2}$)	254	263	265
300% Modulus ($kg\,cm^{-2}$)	148	157	149
Physical properties—220-min cure at 140°C			
Tensile strength ($kg\,cm^{-2}$)	216	248	249
300% Modulus ($kg\,cm^{-2}$)	121	149	144
Aged 10 days at 85°C—after 220-min cure at 140°C			
Tensile strength ($kg\,cm^{-2}$)	170	222	215

cal properties are essentially the same but there is a slight but consistent advantage in reversion resistance for the two MBTS-based systems. These systems are equally applicable in blends of NR with BR, IR or SBR.[16]

TABLE 13

PROCESSING AND CURING CHARACTERISTICS OF MBTS V. MBS IN NR SEMI-EV

	Conventional		Semi-EV	
MBS	—	0·6	—	0·6
MBTS	0·9	—	0·9	—
DTDM	—	—	0·6	0·6
Sulphur	2·5	2·5	1·5	1·5
Processing characteristics, Mooney at 121°C, scorch time t_5 (min)	14	24	20	39
Curing characteristics, rheometer at 141°C				
t_{90} (min)	24	32	20	32
Maximum modulus (in lb)	70	77	81	78

TABLE 14

PHYSICAL PROPERTIES OF MBTS V. MBS IN NR SEMI-EV

	Conventional		Semi-EV	
MBS	—	0·6	—	0·6
MBTS	0·9	—	0·9	—
DTDM	—	—	0·6	0·6
Sulphur	2·5	2·5	1·5	1·5
Physical properties—optimum at 140°C				
Tensile strength (kg cm^{-2})	208	208	212	220
300% Modulus (kg cm^{-2})	200	181	198	200
Elongation at break (%)	310	350	350	350
Cured 220 min at 140°C, retention tensile strength, %	86	82	93	90

Whilst both semi-EV systems give comparable ageing properties, those based on DTDM have superior initial fatigue resistance, and show a good retention of this property on ageing (Fig. 15).

The semi-Ev systems give good processing and cure properties, long-

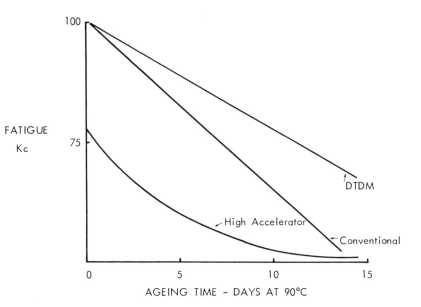

Fig. 15. Semi-EV systems—aged fatigue in NR (fatigue life measured by extension).

est scorch being given by the sulphur donor system and fastest cure by the high accelerator system. Where maximum resistance to reversion and ageing is required, and dynamic properties are of less importance, in, for example, steam hose, heat-resistant cables and high-temperature-cured mechanicals, the fully efficient systems are used.[30,31] Properties for a semi-EV and EV system are compared in Table 15.

Properties for alternative EV systems are given in Tables 16 and 17.

The influence of curing temperature on physical properties for conventional and fully efficient curing systems is shown in Fig. 16. Whilst the full EV compound does suffer some fall in tensile strength as the cure temperature approaches 200°C, it is clearly much superior to the conventional.

Figure 17 summarises the relative performances of the conventional with semi-EV and full EV systems based on DTDM or high accelerator in natural rubber.

Further improvements in the properties of EV systems have been found when soluble compounding ingredients are used such as MBS

TABLE 15

AGEING AND REVERSION RESISTANCE IN SEMI-EV AND EV NR

	Conventional	Semi-EV	Full EV
MBS	0·6	0·6	1·0
DTDM	—	0·6	1·0
TMTD	—	—	1·0
Sulphur	2·5	1·5	—
Physical properties at optimum cure			
Tensile strength ($kg\,cm^{-2}$)	182	172	185
300% Modulus ($kg\,cm^{-2}$)	139	134	140
Elongation at break (%)	400	400	400
Compression set ASTM D 395 22 h at 70°C	35	27	15
Aged 10 days at 90°C, retention tensile strength, %	40	75	88

TABLE 16

REVERSION RESISTANCE IN NR EV

	Conventional	High accelerator	Sulphur donor
MBS	0·5	3·0	1·1
TMTM	—	0·6	—
TMTD	—	—	1·1
DTDM	—	—	1·1
Sulphur	2·5	0·5	—
Cured 2 min at 200°C, retention tensile strength (of 140°C value), %	40	85	77
Cured 15 min at 200°C, retention tensile strength (of 140°C value), %	29	76	72

Longest scorch safety is given with the high accelerator/low sulphur combination (see Table 17).

TABLE 17
PROCESSING AND CURING CHARACTERISTICS IN NR EV

	Conventional	High accelerator	Sulphur donor
MBS	0·5	3·0	1·1
TMTM	—	0·6	—
TMTD	—	—	1·1
DTDM	—	—	1·1
Sulphur	2·5	0·5	—
Processing characteristics, Mooney at 121°C, scorch time t_5 (min)	33	38	21
Curing characteristics, rheometer at 140°C			
t_{90} (min)	31	30	26
$t_{90} - t_2$ (min)	21	14	18

Scorch safety of the sulphur donor system can be increased by reducing the level of TMTD or TMTM and increasing DTDM or MBS.

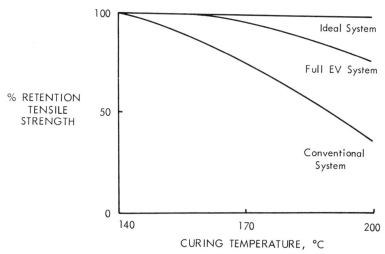

FIG. 16. Effect of curing temperature on tensile strength for NR at optimum cure.

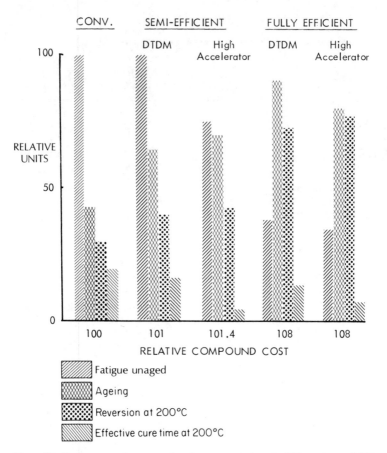

FIG. 17. Summary of properties for conventional, EV and semi-EV curing systems in NR.

rather than other sulphenamides, butyl thiurams rather than methyl thiurams, lauric acid rather than stearic acid and sulphur levels of maximum 0·5 phr.[32,33]

6.1.2. EV and Semi-EV Systems in SBR

In SBR, reductions in elemental sulphur do not reduce fatigue life as in NR; thus the benefits of improved ageing resistance can be achieved without compromising fatigue resistance (Table 18). Semi-EV systems are, however, used in SBR due to the need to compromise on costs, as at low sulphur levels the required accelerator levels are high (Table 10).

TABLE 18

FATIGUE LIFE AT 100% STRAIN[24] IN SBR AND NR

	kc to failure	
	NR	SBR
Conventional	44·5	400
EV	18·0	450

As seen from the crosslink distribution given in Table 9, a so-called NR EV network has apparently a similar crosslink type distribution to a conventional SBR network. The use of the term EV in SBR does not therefore imply the same crosslink network as when the term is applied in NR, although both imply a low sulphur level (from 0 phr to approximately 0·75 phr).

Oxidative ageing of conventional-cured SBR results in dramatic increases in both hardness and modulus with corresponding reduction in elongation at break. In SBR, the EV and semi-EV curing systems result in lower aged modulus and hardness, better retention of elongation at break and general overall improvements in compression set and heat build-up.[25,27,34,35]

Essentially the same approach is used in SBR as in NR to match modulus as elemental sulphur level is reduced. Accelerator levels are, however, higher. It is possible to reduce these by the use of activators (see Table 19).

TABLE 19

EFFICIENT VULCANISING SYSTEMS IN SBR[a]

Curing systems	Conventional	Semi-EV		Full EV		
CBS	1·2	2·5	1·0	7·0	4·0	1·2
DTDM	—	—	1·0	—	—	1·2
TMTM	—	—	—	—	1·0	—
TMTD	—	—	—	—	—	1·2
Sulphur	2·0	1·2	1·2	0·75	0·75	—

[a] Belting compound: SBR-1712 137·5, HAF black 50·0, zinc oxide 5·0, stearic acid 1·0.

For semi-EVs high accelerator/low sulphur levels are more economical than using DTDM. However, the converse is true in the case of full EVs. The increase in ageing resistance obtained with semi-EV systems is not as marked as in NR. As can be seen in Table 20, the two semi-EVs do reduce marching modulus but by a relatively small amount.

TABLE 20

REDUCTION OF MARCHING MODULUS FOR SEMI-EV IN SBR

	Conventional	Semi-EV	
CBS	1·2	2·5	1·0
DTDM	—	—	1·0
Sulphur	2·0	1·2	1·2
Original 200% modulus (kg cm^{-2})	70	67	66
Aged 6 days at 85°C, increase in 200% modulus (%)	120	90	100

Table 21 compares fatigue properties of the semi-EVs with the conventional control. There is a tendency for initial fatigue resistance to improve with cure efficiency in contrast to NR. Aged fatigue properties are also improved as expected.

TABLE 21

FATIGUE PERFORMANCE OF SEMI-EV IN SBR

	Conventional	Semi-EV	
CBS	1·2	2·5	1·0
DTDM	—	—	1·0
Sulphur	2·0	1·2	1·2
Fatigue life—80% extension kc to failure	323	275	416
Aged 3 days at 85°C, fatigue life—80% extension, kc to failure	61	81	117

The improvement in ageing properties when full EV systems are employed can clearly be seen in Table 22. Fatigue properties, both unaged and aged, are superior to the conventional compound. This is in complete contrast to NR where the full EV systems give very poor fatigue properties.

TABLE 22

PROPERTIES OF FULL EV SYSTEMS IN SBR

	Conventional	Full EV	
MBS	1·2	7·0	1·2
TMTD	—	—	1·2
DTDM	—	—	1·2
Sulphur	2·0	0·75	—
Aged 10 days at 90°C, % increase 300% modulus	115	65	60
Compression set, 22 h at 70°C, %	18	11	11
Fatigue—75% extension			
Unaged—kc	645	750	800
Aged—kc	192	445	400

Semi-EV systems give similar processing properties to the conventional, but with faster rates of cure. Scorch times for the full EVs are lower than the control, particularly so for the DTDM system. It is, however, possible to adjust scorch properties by using lower levels of activator and increasing sulphenamide or DTDM levels (Table 23).

The overall benefits in oxidative-ageing resistance produced by reducing sulphur in SBR are substantial. The choice for a particular application depends on a combination of factors including processing and curing characteristics, cost and level of performance required. The semi-EV system with about 1·0 phr elemental sulphur gives a good balance of aged and unaged properties with excellent curing characteristics. For the greatest improvements in oxidation resistance, however, a sulphurless system is required. The level of thiuram should, however, be kept to a minimum due to the potentially deleterious effect on

TABLE 23

PROCESSING AND CURING CHARACTERISTICS

	Conventional	Semi-EV		Full EV		
CBS	1.0	2.5	1.0	7.0	4.0	1.2
DTDM	—	—	1.0	—	—	1.2
TMTM	—	—	—	—	1.0	—
TMTD	—	—	—	—	—	1.2
Sulphur	2.0	1.2	1.2	0.75	0.75	—
Processing characteristics, Mooney at 135°C, scorch time t_5 (min)	22	17	20	19	15.2	10.0
Curing characteristics, rheometer at 153°C, $t_{90} - t_2$ (min)	12.6	7.6	9.3	8	7.0	10.0

fatigue.[20] The exact reasons for this are not clear but there are references[34,37] to agglomeration of compounded ingredients as a cause of cracking due to their serving as flaw initiation points.

6.1.3. EV and Semi-EV Systems in NBR

EV systems have been used for some time in NBR to obtain good ageing resistance and in particular for oil seals and similar compounds, low compression set. A comparison of conventional and EV systems in NBR[38] is shown in Table 24.

TABLE 24

CONVENTIONAL AND EV SYSTEMS IN NBR

	Conventional		EV	
MC Sulphur	1.5	0.5	0.25	—
TMTD	0.1	1.0	2.5	4.0
MBTS	1.0	0.5	1.5	1.0
Compression set, 22 h at 100°C, %	53	35	24	23
Retained elongation at break, Aged 70 h at 100°C, %	53	58	74	70
Bloom	none	slight	medium	heavy

As in SBR, oxidative-ageing resistance progressively improves with lower elemental sulphur; fatigue life is unaffected. The sulphurless high-TMTD EV system has been used widely in NBR but gives problems due to low processing safety and problems with the heavy crystalline bloom on the surface of cured and uncured compounds. Processing safety can be increased and bloom problems minimised by reductions in TMTD level and substitution by a sulphenamide or sulphur donor (Table 25).

TABLE 25

EV SYSTEMS IN NBR

TMTD	4·0	1·5
MBTS	1·0	—
DTDM	—	1·0
Processing characteristics, Mooney at 121°C, scorch time t_5 (min)	15	20
Curing characteristics, rheometer at 153°C, cure time t_{90} (min)	20	22
Bloom	heavy	none
Compression set, 22 h at 100°C, %	23	27

Uniform dispersion of sulphur in NBR can be difficult in spite of the use of magnesium carbonate-coated sulphur (MC sulphur). The use of lower elemental-sulphur levels and their partial or complete replacement by accelerators or sulphur donors is an obvious solution to this problem. Where compression set is particularly critical the use of sulphur donors or alternative accelerators to TMTD must be evaluated with care as it may tend to increase at lower TMTD levels. Figure 18 illustrates the use of full EV systems in NBR based on different accelerators. Two fully efficient vulcanising systems based on DTDM have been compared with the commonly used high accelerator/low sulphur combination with respect to ageing over a wide range of temperatures. The graph clearly shows the superiority of the MBS/TMTD/DTDM combination over the other systems.

6.1.4. EV and Semi-EV Systems in EPDM and Other Polymers

The principles of efficient vulcanisation are equally applicable to EPDM and other sulphur-vulcanisable polymers. From the starting point of conventional sulphur levels (for EPDM approximately 1·5 phr) progres-

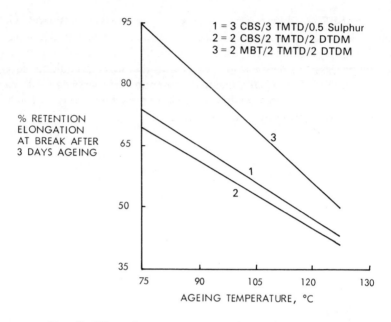

Fig. 18. Effect of temperature on ageing performance in NBR.

sive replacement by accelerators and/or sulphur donors will give improved ageing and compression set and unaltered fatigue resistance. Curing systems for EPDM are normally based on thiazoles with dithiocarbamates and thiurams. Because of the low degree of unsaturation in EPDM compared with other polymers, high accelerator loadings (2–10 phr) comprising different types of accelerators are frequently used to achieve sufficiently fast cure rates and high modulus. The low solubility of the thiurams and dithiocarbamates usually means that a number of individual accelerators must be added at low levels (<1·0 phr). Sulphenamide accelerators can be used in EPDM where their high solubility is a distinct advantage (Table 26).

A comparison of MBT, MBTS and CBS as primary accelerators shows that their relative effect on cure rate is in direct contrast to that which might be expected in other general-purpose polymers with CBS giving a 30% reduction in cure time compared to MBTS.

Other non-blooming accelerators such as the thiophosphates (e.g. ZBDP) are also widely used in EPDM.[39,40]

TABLE 26

COMPARISON OF THIAZOLE AND SULPHENAMIDE ACCELERATORS IN EPDM

MBT	1·5	—	—
MBTS	—	1·5	—
CBS	—	—	1·2
TMTD	0·8	0·8	0·7
Sulphur	1·5	1·5	1·5
Mooney scorch at 135°C t_5 (min)	7·4	9·2	8·6
Rheometer at 160°C			
t_{90} (min)	21	19	13·3
Maximum modulus (in lb)	60	62	60

7. HIGH-TEMPERATURE CURING

Current trends towards production methods with higher outputs have in common an inevitable move to higher curing temperatures. Curing by continuous vulcanisation (CV) or by injection moulding involves temperatures appreciably higher (180–240°C) than those used in more traditional press and autoclave cures (140–150°C). Similarly in press curing, substantial improvements in output can be made by increasing curing temperatures to reduce mould opening times. As the temperature of cure increases, cure time is usually reduced to maintain the same equivalent cure input. The exact isothermal cure time for a given compound at a given temperature can be determined by cure meter but the reduction in cure time is of the order of 50% for each 10°C rise in temperature. However, although the state of cure achieved may well be optimum for the particular compound and temperature chosen, the actual level of physical properties obtained is strongly dependent on the actual cure temperature. As curing temperature increases, physical properties (tensile strength, tear strength, modulus, elongation at break, hardness and resilience) deteriorate. This can be correlated with reduced crosslink density in the vulcanisate and is illustrated in Fig. 19 for a conventionally cured natural rubber compound cured in 20°C steps from 140°C to 200°C.

When curing temperatures are increased, the efficiency of the crosslink reaction is reduced and the optimum crosslink density decreases. At cure times well beyond optimum, crosslink density values plateau out,

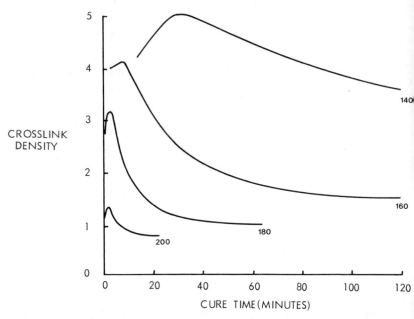

FIG. 19. Influence of the curing temperature on crosslink density in NR.[44] Sulphur 2·5, CBS 0·6.

particularly at temperatures in excess of 160°C. Consequently, cure temperature is more important than cure time in determining the crosslink density achieved for compounds which are overcured, as, for example, the surface of a thick conveyor belt. Although the amount of sulphur combined in the network remains essentially constant, as cure temperature increases, the efficiency of sulphur utilisation falls (Table 27).

As the amount of total combined sulphur (S_c) remains approximately constant and E values increase, this means that crosslink density decreases (see definition of E parameter, Section 3).

As the $(E' - 1)$ values increase correspondingly with cure temperature, this shows that more and more sulphur is used in forming main-chain modifications[42] instead of crosslinks. At cure temperatures of 160°C and above the average length of the crosslinks is very short, being mainly monosulphidic. The trend to shorter, and ultimately monosulphidic, crosslinks at the higher cure temperatures can be clearly seen in Fig. 20. In a conventional NR compound cured 2 h at 140°C, the majority of crosslinks are polysulphidic. However, after 20 min cure at 180°C only di- and a preponderance of monosulphidic crosslinks are found.

TABLE 27

EFFICIENCY OF CROSSLINKING REACTIONS V. CURING TEMPERATURE FOR NR GUM

Curing temperature, °C	140	160	180
Cure time, min	360	120	60
S_c	5·4	5·47	5·44
E	27·0	37·7	50·4
$E' - 1$	24·2	36·5	49·0
$E - (E' - 1)$	3	1	1

S_c: Total combined sulphur × 10^4 gram atoms/gram rubber.
E and $E' - 1$: see definitions in Section 3.
$E - (E' - 1)$: number of gram atoms of sulphur in crosslink/chemical crosslink.

When compounds are overcured at high temperatures it can be concluded that

1. crosslink density is drastically reduced;
2. essentially monosulphidic crosslinks are obtained;
3. there is a large increase in main-chain modifications.

Of these factors, the most important in determining initial physical properties is the crosslink density.

Reductions in physical properties are most pronounced in NR or NR-containing blends but when conditions are particularly severe, it can occur with SBR. The example is shown in Table 28 for an SBR/BR

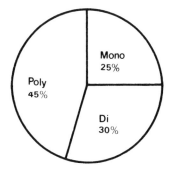
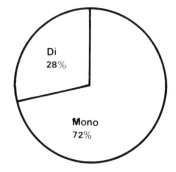

FIG. 20. Crosslink distribution in NR. Sulphur 2·5, CBS 0·6. Left, cure 120 min at 140°C; right, cure 20 min at 180°C.

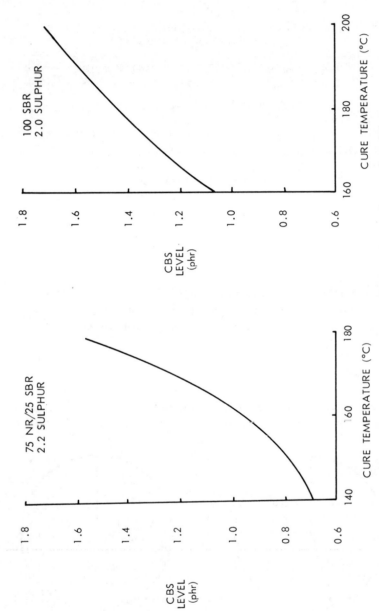

FIG. 21. Optimium accelerator level at a given temperature for equal modulus.

blend where properties of samples cured at 170°C are compared with samples cured at 190°C and 205°C but for longer than optimum cure times (approximately 3 and 5 times optimum cure). The reduction in crosslink density with increased cure input is much smaller than with NR compounds.

TABLE 28

INFLUENCE OF THE CURING TEMPERATURE ON THE CROSSLINK DENSITY IN SBR/BR

Sulphur (phr)	2	2	2
CBS (phr)	1	1	1
Curing temperature (°C)	170	190	205
Cure time (min)	20	15	10
Crosslink density $(2M_c)^{-1} \times 10^5$	5	4·3	4·1

EV and semi-EV systems have proved to be of particular value because of their high rates of cure and effect in suppressing reversion in NR.[33] Studies over a range of cure temperatures from 140°C to 220°C for tyre-tread compounds based on NR, SBR and blends with BR showed that a vulcanisation system based on 1·5 phr sulphur and 3·5 phr CBS gave the best results over the range of temperature, sulphur and CBS levels examined.[43] However, changing to a semi-EV or EV system, whilst reducing the drop in physical properties at high temperature, does still present some decrease and an alternative solution has been to increase accelerator levels whilst maintaining elemental sulphur at conventional levels.[41,44] A relationship was developed to give the optimum level of sulphenamide for a given cure temperature to maintain physical properties (Fig. 21).

Figure 21 shows for 75/25 NR/SBR and 100 SBR that modulus values can remain constant over the full range of curing temperatures using the recommended increased sulphenamide levels.

8. NON-SULPHUR VULCANISATION

Workers at MRPRA have proposed urethanes as an alternative form of crosslinking to that based on sulphur bridges[45] and vulcanising chemicals based on such products are commercially available. The vulcanising agent in these systems is derived from a *p*-benzoquinone monoxime

(p-nitrophenol) and a di- or polyisocyanate. Additives as used in sulphur vulcanisation are not necessary, but the efficiency of the process is improved by the presence of free diisocyanate and by ZDMC. The latter catalyses the reaction between the nitrosophenol and the polymer chain to form pendent groups.

The principal advantage of these systems lies in the high stability of the crosslinks which give very little modulus reversion even on extreme overcure. Problems can occur with their lower scorch, rate of cure and modulus. However, modulus and fatigue-life retention on ageing is very good. Work in a number of laboratories is aimed at seeking crosslink systems which will be thermally labile at high temperatures, but perform elastically at operating temperatures, thus bringing rubber moulding closer to plastics technology. One such patent[46] uses an elastomer obtained by reacting a metal salt with a coordinating basic group present in an elastomer containing an electron-donating atom. Copolymers of BR, SBR and vinylpyridine may be used with zinc, nickel and cobalt chlorides.

REFERENCES

1. Hu, P. L. and Scheele, W. *Kautsch. Gummi Kunstst.*, **15**, 1962, 440.
2. Campbell, R. H. and Wise R. W. *Rubber Chem. Technol.*, **37**, 1964, 635.
3. Dibbo, A., Lloyd, D. G. and Payne, J. *Rubber Chem. Technol.*, **45**, 1972, 1513.
4. Skinner, T. D. *Rubber Chem. Technol.*, **45**, 1972, 182.
5. Armstrong, R. T., Little, J. R. and Doak, K. W. *Rubber Chem. Technol.*, **17**, 1944, 788.
6. Decker, G. E., Wise, R. W. and Guerry, D. *Rubber Chem. Technol.*, **36**, 1963, 451.
7. Lloyd, D. G. *J. Rubber Res. Inst. Malaya*, **22** (4), 1969, 399.
8. Bueche, F. *J. Polymer Sci.*, **24**, 1958, 189.
9. Bueche, F. *J. Polymer Sci.*, **33**, 1958, 259.
10. Studebaker, M. L. *Rubber Chem. Technol.*, **39**, 1966, 785.
11. Kuvshinskii, E. V. and Sidorovich, E. A. *Rubber Chem. Technol.*, **32**, 1959, 662.
12. Saville, B. and Watson, A. A. *Rubber Chem. Technol.*, **40**, 1967, 100.
13. Hammersley, D. A. H., Lloyd, D. G. L., Neale, A. J. and Rodger, E. R *Intern. Rubber Conf., Moscow*, 1969.
14. Westlinning, H. *Rubber Chem. Technol.*, **43**, 1970, 1194.
15. Mastromattes, R. P., Mitchell, J. M. and Brett, T. J. *Rubber Chem Technol.*, **44**, 1971, 1065.
16. Taylor, R. D. *ACS Div. Rubber Chem., Toronto*, May 1974.
17. Leib, R. I., Sullivan, A. B. and Trivett, C. D. *Rubber Chem. Technol.*, **43**, 1970, 1188.

18. LEYLAND, B. N. and MEYRICK, T. J. *Intern. Rubber Conf. Munich*, September 1974.
19. ROEBUCK, H. *European Rubber J.*, **155** (8), 1973, 10.
20. DILHOEFER, J. R. and RODGER, E. R. *3rd Australian Rubber Technol. Conf., Terrigal*, September 1974.
21. KEMPERMANN, Th. and REDETZKY, W. *Kautsch. Gummi Kunstst.*, **22**, 1969, 706.
22. DILHOEFER, J. and PAYNE, J. *1st Australian Rubber Technol. Conf., Terrigal*, September 1968.
23. SWEENEY, T. *Rubber World*, **56**, 1971.
24. STUDEBAKER, M. L. *Rubber Chem. Technol.*, **39**, 1966, 1526.
25. STUDEBAKER, M. L. and BEATTY, J. R. *Rubber Chem. Technol.*, **45**, 1972, 450.
26. LEE, T. C. and MORRELL, S. H. *J. Inst. Rubber Ind.*, **7**, 1973, 27.
27. BUSWELL, A. G. *J. Inst. Rubber Ind.*, **8**, 1974, 28.
28. AYERST, R. C., LLOYD, D. G. and RODGER, E. R. *Kautsch. Gummi Kunstst.*, **11**, 1971, 583.
29. MCCALL, E. B. *J. Rubber Res. Inst. Malaya*, **22** (3), 1969, 354.
30. SKINNER, T. D. and WATSON, A. A. *Rubber Age (NY)*, November/December 1967.
31. LEYLAND, B. N. and MEYRICK, T. J. *1st Australian Rubber Technol. Conf., Terrigal*, September 1968.
32. PORTER, M. *NR Technol.*, **4** (4), 1973, 76.
33. WHEELANS, M. A. *2nd Ann. Nat. Conf. IRI, Blackpool*, May 1974.
34. RAILSBACK, H. E., HOWARD, W. S. and STUMPEN, N. A. JR.. *Rubber Age*, **46**, 1974, 106.
35. COX, W. L. *ACS Div. Rubber Chem., Boston, Mass.*, April 1972.
36. SMITH, R. W. and BLACK, A. W. *Rubber Chem. Technol.*, **37**, 1964, 338.
37. HESS, W. M. and BURGESS, K. A. *Rubber Chem. Technol.*, **36**, 1963, 754.
38. *Curing Systems for Nitrile Rubber*, Monsanto Co. Technical Bulletin O/RC-16.
39. *Curing Systems for EPDM*, Monsanto Co. Technical Report No. 8A.
40. *A General Purpose Vulcanisation System for EPDM*, Monsanto Co. Technical Note No. 107/78.
41. DAVIES, K. M. *Intern. Rubber Conf., Brighton*, 1977, paper 49.
42. LOO, C. T. *Polymer*, **15**, 1974, 357.
43. BLOW, C. M. and LOO, C. T. *J. Inst. Rubber Ind.*, **7**, 1973, 205.
44. *Increased Tyre Productivity—Part III: Accelerator Systems for High Temperature Curing*, Monsanto Co. Technical Bulletin IC/RC-31.
45. BAKER, C. S. L., BARNARD, D. and PORTER, M. *Rubber Chem. Technol.*, **43**, 1970, 501.
46. Aquitaine Total Organico, GB 1,339,653.

Chapter 4

CARBON BLACKS

A. I. MEDALIA and R. R. JUENGEL

Cabot Corporation, Billerica, Massachusetts, USA

and

J. M. COLLINS

Cabot Carbon Limited, Stanlow, Ellesmere Port, UK

SUMMARY

Carbon black is examined in terms of its physical form, characterisation and mechanism of reinforcement of polymers. The characteristics of 'new technology' blacks are indicated as distinct from conventional grades. Production and quality-control aspects are considered and the selection of carbon black for practical applications is reviewed. Finally, possible future changes in the industry are postulated.

1. BASICS OF CARBON BLACK

1.1. Introduction: Formation of Carbon Black

Carbon black is the prime reinforcing agent for rubber. Most rubber articles produced are compounded with carbon black, typically in the amount of at least 50 parts of black per hundred parts of rubber by weight (50 phr). Reinforcement by carbon black involves an increase in the resistance of the rubber to abrasion, tearing, and other types of tensile failure ('ultimate properties'); and an increase in the hardness, modulus of elasticity, and related viscoelastic properties. While these effects are most dramatic with a non-crystallising rubber, such as SBR, which lacks inherent strength, they were first found and put into practice with a crystallising rubber (NR). The discovery of reinforcement by carbon black, developed in England in the early 1900s, is considered to rank just after that of vulcanisation in importance to the rubber industry. The years since then have witnessed the development of many grades of carbon black tailored to the special needs of the rubber and other industries; as well as the development of many other fillers, some of which are also reinforcing, but none to the degree exhibited by the higher-quality grades of carbon black.

Carbon black is a colloidal form of elemental carbon. It owes its reinforcing character to its colloidal morphology (the size and shape of the ultimate units) and to its surface properties. In order to understand these aspects and the manner in which they are controlled by the carbon-black manufacturer, it is helpful to have a picture of the way in which carbon black is formed. At the present time most carbon black is produced by the oil-furnace process (furnace black). In this process a hydrocarbon of high BTU content, such as natural gas or naphtha, is burned in a specially designed furnace, and a highly aromatic oil ('feedstock') is sprayed into the furnace, where the feedstock cracks to form carbon black. In detail, it appears that the feedstock droplets vaporise and undergo partial molecular rearrangement and decomposition. Large polyaromatic molecules are formed, which pass through various stages of nucleation and fusion, resulting in the formation of spherical particles. As the polyaromatic molecules lose hydrogen, the particles pass from a liquid state, through a viscous state, to a solid carbon state. In the liquid state, collision of two particles (droplets) produces a single larger spherical particle. In the viscous state, particles adhere to each other upon collision, and partially fuse together over a small area of contact. Collision of many particles in the viscous state results in fused *aggregates* which are the primary units of carbon black (Fig. 1a). Further dehydrogenation in the furnace hardens the aggregates and render them incapable of fusion with other aggregates, thus limiting the aggregate size; however, aggregates can be held together in *agglomerates* by weak secondary forces.

The entire process of carbon-black formation takes place over a period of milliseconds at temperatures of 1100–1800°C. Control of the colloidal morphology involves primarily control of the relative length of time the particles are in the liquid and viscous states, and the concentration and rates of collision of the particles in these states. This is achieved by furnace design; by control of flow rates of air, gas, and feedstock; and by use of trace additives, especially alkaline metal salts which affect the particle charge and thus affect the rate of collision. A

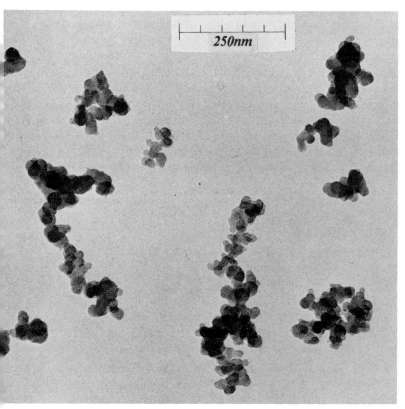

Fig. 1a. Electron micrograph of furnace black.

the end of the carbon-black-formation process, the reaction is *quenched* by a spray of water. If the reaction is quenched too soon, an appreciable fraction of the polyaromatic molecules are not sufficiently converted to carbon and are still extractable as tars; while if it is quenched too late, the surface of the aggregates may be annealed (incipient graphitisation) and lose some reinforcing character. Variations in the design of the furnace and quenching system can affect the time scale of the processes involved and thus affect the *distribution* of aggregate size and surface properties.

After formation and quenching, the carbon-black aggregates are separated from the combustion gas stream by, for example, filtration through fabric bags.

1.2. Nature of Carbon Black

With an aggregate of carbon black, the carbon atoms are arranged in graphitic (i.e. polyaromatic) layers. The layers are generally parallel to the surface, with the inner layers being generally concentric (Fig. 1b).[3] Over small regions the layers are parallel to each other but are not in the same orientation as in graphite. At the surface the layers may be somewhat shingled, exposing edge atoms which, together with defects within the layer, appear to act as high-energy sites for adsorption of rubber molecules. Such sites are said to impart high 'surface activity'. If a carbon black is heated in the range 1000–2700°C, the layers are progressively annealed, removing defects and edge atoms and enlarging the

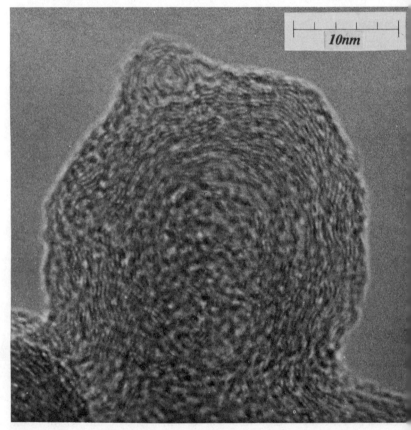

FIG. 1b. High resolution micrograph of carbon-black particles.

regions of parallelism. A carbon black heated to 2700°C is called 'graphitised', although the layers are still not fully orientated. Graphitised blacks have a very homogeneous surface with virtually no high-energy sites. While not commercially available, graphitised blacks have been useful in surface chemistry investigations and in studying the mechanism of reinforcement.[4]

Carbon blacks, as produced, contain appreciable amounts of hydrogen, oxygen, and (depending on the feedstock) sulphur. The sulphur is mainly present as a tightly bound heterocyclic substituent for carbon in the polyaromatic layers, and appears to have little if any influence on the properties. The hydrogen is bonded to edge and defect carbon atoms, both at the surface and in the interior, and to surface oxygen atoms. Oxygen is present at the surface, where it is bonded to edge and defect carbon atoms, chiefly in the form of quinone dioxine and phenolic groups.[5] Furnace blacks contain typically around 1% oxygen by weight (per 100 m^2 of surface area). This can be driven off, in the form of CO (along with a little CO_2 from carboxyl and lactone groups), by heating to about 1000°C. The loss in weight at 950°C is called the *volatile* content of the carbon black. Blacks made by the old 'channel' process contain about 2–3% oxygen (per 100 m^2), which appears to be the limiting amount of oxygen which can be put on the surface. Oxidation of furnace blacks by strong oxidants such as ozone or nitric acid below about 200°C increases their oxygen content to the same level as a channel black. Such blacks are useful in special applications, such as where slow curing is desired.[6] Oxidation at higher temperatures, for example by oxygen, steam, or CO_2, burns away layers and forms slit-like pores.

The most important single property of carbon black is its external specific surface area, i.e. the area accessible to rubber molecules per unit weight of carbon black. High area is associated with a high level of reinforcement; but at the expense of high cost, high hysteresis, and more-difficult processing.

The balance between these properties is one of the chief factors determining the grade of carbon black. For non-porous spheres, the specific surface area is inversely proportional to the particular size (diameter). In carbon-black aggregates the particles which were created in the furnace are still apparent as nodules. Note (Fig. 1b) the continuity of the carbon layers between nodules. The size of the nodules can be assessed under the electron microscope but the measurement is subjective and imprecise. Recent attempts to estimate particle size from image analysis[7] or from tinting strength[8] have been based on correlations with electron-microscope measurements, for a limited group of blacks. It is possible that further improvements in image analysis may permit a direct, automated measurement of particle and aggregate size and distribution, but always subject to the choice of an algorithm for defining particle size. Grade-classification schemes, such as that of the ASTM,[9] which rely on particle size, have led to considerable difficulty and should probably be replaced by surface-area criteria.

The second most important property is the so-called 'structure'. This refers to the bulkiness of the carbon-black aggregates. In general[10] high bulkiness is associated with a large average number of particles per aggregate, N_p. However, the relation between bulkiness and N_p depends to some extent on the manufacturing process,[11] due to variations in the arrangement of particles in the aggregate. In an 'open' aggregate the particles are arranged in a highly branched configuration, so that high bulkiness is achieved with relatively few particles (Fig. 2). In a 'clustered' aggregate the particles are grouped tightly together, so that with the same N_p, the bulkiness is lower. However, since aggregates become more open as they grow, a 'clustered' aggregate of high N_p will have the same bulkiness as an 'open' aggregate of small N_p. The ability to vary the 'clustering' or 'openness' of aggregates is one of the important recent developments in the technology of carbon-black production (see Section 2).

1.3. Characterisation of Carbon Blacks

Carbon blacks for rubber use are generally produced to meet quality specifications, either as established by standards organisations,[9] or as agreed on directly between user and supplier. The specifications are written in terms of readily measurable properties of the carbon black itself and as compounded in standard rubber compounds. However, in view of its complex colloidal nature, a sample of carbon black cannot be described completely by such tests.[12]

From the above discussion it is evident that products of different manufacturers, made in different furnaces of patented design, with different feedstocks, operating conditions, and pelletisation techniques, may differ significantly even though meeting the same basic control specifications and providing approximately equivalent product performance. For this reason it is important for the user to be aware of the meaning of the tests used to characterise carbon black in research as well as in quality control (Section 3). Due to space limitations, only the major testing features can be discussed here.

Fig. 2. Top left, Low 'structure' (low bulkiness, DBPA), low N_p (28), compact morphology. Top right, High 'structure' (high bulkiness, DBPA), low N_p (28), open morphology. Bottom, High 'structure' (high bulkiness, DBPA), high N_p (84), intermediate morphology.

In colloid and surface science, surface area is preferably measured by adsorption of an inert gas, such as nitrogen. Adsorption data obtained at several partial pressures from about 0·05 to 0·3 generally give a straight-line plot in the BET equation.[13] From the slope and intercept the amount of nitrogen required to cover the surface depth of one molecule (a monolayer) can be calculated. This is divided by the effective surface area per molecule (the 'squat' area, generally taken as 16·2 Å² for nitrogen) to give the surface area.

The BET procedure is time-consuming and also fails to distinguish between external and internal surface area. For ease and rapidity, it can be simplified by making the adsorption measurement at a single partial pressure, with the assumption of a constant or zero intercept (reasonably valid for most carbon blacks). Several instruments are commercially available for carrying out this well-recognised single-point procedure.[14]

The simplest measurement related to surface area is the ASTM Iodine Number.[15] This test is based on adsorption of iodine from aqueous potassium iodine solution. It is rapid and precise but unfortunately is affected somewhat by, for example, the presence of residual extractable materials and surface oxygen.[16] The test conditions were originally designed so that the number of milligrams of iodine adsorbed per gram (the 'iodine number') would be equivalent to the BET area. However with blacks made by different processes, small systematic deviations from this relation are apparent.[11]

Internal surface area, in the form of slit-like pores, can be developed by oxidation of carbon black at elevated temperatures. These pores are inaccessible to rubber molecules and thus do not contribute to reinforcement.[17] One method of distinguishing internal from external areas is to fill the pores with nitrogen at low partial pressure, and then measure the amount of nitrogen required for the second and third layers. This so called 't' method[18,19] is accurate but exceedingly time consuming. A more rapid method involves the adsorption of a large molecule such as cetyl trimethylammonium bromide (CTAB) from aqueous solution.[20] This has been shown to measure only external area without significant interference from extractables or surface-absorbed oxygen.[16] The principal drawback of the method is the necessity of separating the carbon black from the CTAB solution before back titrating the CTAB. Since CTAB is an excellent surfactant, a fine colloidal dispersion is formed, and recourse must be made to high-speed centrifugation or ultrafiltration. Porosity is generally absent in rubber grades of carbon black[19] but some may be present with grades of very high surface area (120 m^2/g) or if the black has been oxidised through, for example, improper handling in the drier.

'Structure' is assessed by addition of a liquid which wets the black such as dibutylphthalate (DBP), until the crumbly mass of dry black suddenly coheres. This is measured using a machine which senses the end point as a sharp rise in viscosity.[21] The average bulkiness of the individual aggregates can be calculated by subtracting the amount of DBP trapped *between* aggregates at the endpoint.[10] 'Structure' can also be estimated from the bulk density of the dry black, either as pellets or under a pressure of 1000–5000 psi in a compression cylinder.[22]

The energy required for the incorporation of carbon black in rubber fractures some aggregates and thus reduces the 'structure'. In order to bring the black to this lower 'structure' level before running a DBP absorption test, a sequence of four similar compression steps has been proposed.[23] The degree of breakdown resulting from this '24M4' procedure is generally more severe than during incorporation in rubber,[7] depending on the type of rubber, grade of black, and incorporation procedure. While the crushing procedure makes the test longer and less sensitive, it may be useful in distinguishing blacks which differ in their degree of breakdown under mechanical stress.

The amount of carbon per aggregate is an important property which may be related to reinforcement and which, taken in conjunction with surface area and 'structure', gives a measure of clustering versus openness. A simple optical measurement, the tinting strength, measures the light absorption coefficient, which is related to the average amount of carbon per aggregate.[24] Higher tinting strength arises from less carbon per aggregate, which can be due to higher surface area (smaller particle size) or fewer particles per aggregate. Aggregate size distribution can be measured by centrifugal sedimentation or by electron microscopy with image analysis.

The surface oxygen groups can be analysed in detail[5] but more commonly one has recourse to the volatile content (see above) and to the pH of an aqueous slurry. Surface activity is estimated from bound rubber measurements[25] and from adsorption of small molecules such as propane, butane, or water.[26]

In recent years, arising from the suspected carcinogenicity of certain polyaromatic hydrocarbons (PAHs), the American Governmental Health Authorities are re-examining the PAHs which are present in the extractable surface residuals of carbon blacks (Section 1.1). The concentration of PAHs normally present is less than 0.1%, and since they are strongly adsorbed on or within the black aggregates, it is necessary to resort to prolonged extraction with high-boiling organic solvents to remove them more or less completely. To date, there is no evidence that carbon blacks themselves are carcinogenic under normal conditions of manufacture or use. Once extracted, very sensitive chromatographic methods are required to detect these PAHs which are present only in concentrations of parts per billion (10^9) and the majority of which are recognised as non-carcinogenic.

1.4. Mechanism of Reinforcement

Of the various aspects of reinforcement, the viscoelastic properties are most readily accounted for. Over 70 years ago Einstein[27] explained the increase in viscosity of a liquid due to the addition of spherical particles, as being due to the perturbation of flow caused by rotation of the particles as the liquid is sheared. Forty years later this treatment was extended to modulus[28] by substituting strain for rate of strain. These theories have been substantiated by experimental measurements with dilute suspensions of spherical or nearly spherical particles, including thermal blacks (Section 3).[29]

In the Einstein treatment, which is strictly valid only at very low concentrations, interaction between particles is neglected, and the increase in viscosity or modulus is proportional to the first power of the volume fraction of particles, ϕ. At higher concentrations, particle interaction leads to higher-order terms. Of the many theoretical equations which have been derived,[29] we may mention particularly that of Guth and Gold:[30]

$$M_f/M_g = 1 + 2\cdot5\phi + 14\cdot1\phi^2 \tag{1}$$

where M_f and M_g are the moduli (or viscosities) of filled and gum compounds, respectively.

Reinforcing blacks give a larger increase in modulus than predicted by eqn. (1). This is because they consist of asymmetric aggregates rather than spheres. Within each aggregate, rubber is occluded and partially shielded from deformation.[10] Thus the effective volume fraction of filler is increased in relation to the extent to which the occluded rubber is effectively immobilised. The volume of occluded rubber can be taken as equal to the volume of DBP which fills the aggregates in a DBP absorption test;[10] a similar treatment can also be applied to crushed DBP.[31] With reinforcing blacks, the occluded volume amounts to between 0·5 and 1·5 times the volume of carbon, depending on 'structure'.

Given a certain calculated volume of occluded rubber, an effectiveness factor can be calculated from an experimentally measured viscoelastic property, such as the low-strain equilibrium modulus.[32] The effective volume fraction of filler is calculated by an appropriate equation such as eqn. (1), and the difference between this and the actual volume fraction is attributed to the effectively immobilised rubber. This treatment has given an effectiveness factor of 0·5 for many blacks in low-strain equilibrium modulus[32] and certain dynamic modulus[33] measurements, i.e. the occluded rubber acts as if half of it were part of the filler. If the volume of occluded rubber is calculated from electron-microscopic measurements after incorporation in rubber, somewhat higher effectiveness factors are found.[7]

Another important viscoelastic property is hysteresis, discussed in more detail in Section 4. The simplest measure of hysteresis is tan δ, which is the fraction of mechanical energy input which is converted to heat in a cyclic strain. For a wide range of blacks, tan δ is a function of surface area and volume fraction.[34,35] This is in accordance with the concept[36] that hysteresis arises from breaking and re-forming of inter-aggregate contacts. At normal loadings, the carbon-black aggregates are in contact with each other, forming a through-going network, as revealed especially by electrical conductivity. Deformation breaks the networks, but its reformation during cycling requires time, and is thus a hysteretic process.

To account for the ultimate properties, theories have developed along two main lines: hysteretic and molecular. As an example of the former, Lake and Thomas[37] point out that in the absence of hysteresis, the only energy required for tear propagation is that needed to fracture molecular bonds and create new surface. With hysteresis, considerable additional energy must be put into the system.

Molecular theories of ultimate properties have considered that filler particles provide a means of stress relief. In Bueche's mechanism,[38] carbon-black aggregates are linked by rubber molecules of different chain lengths. When the sample is stretched to the point where the shortest molecules break, the longer ones carry the load. In the molecular slippage mechanism[39] it is assumed that rubber molecules are not rigidly attached to the carbon black but can slip along the surface. On stretching, this slippage allows for stress relief and prevents bond breakage. Actual detachment of the rubber from the carbon-black surface, with the formation of vacuoles, has been proposed with low-area (semi-reinforcing) blacks,[40] but as a source of weakness rather than strength.

One of the most important (and complex) ultimate properties is treadwear. For most blacks there is a general relation between treadwear and hysteresis; that is, a black which imparts better treadwear in a given recipe also imparts higher hysteresis. This is readily understood on the basis of the hysteretic theories of ultimate properties. However, indiscriminate application of these theories does not seem justified, since hysteresis is generally measured at much smaller deformations (around 10%) than probably occur in treadwear. The molecular theories predict a dependence of ultimate properties on surface area, which as we have seen is also the dominant factor affecting hysteresis. Despite the theories, or perhaps because of their inadequacies, the relation between hysteresis and treadwear does not appear to be immutable, and new blacks are being developed which permit a few per cent improvement in treadwear at a given hysteresis (or conversely) (Section 4.2).

2. 'NEW TECHNOLOGY' AND 'IMPROVED' BLACKS

Around 1970 the grades used for tyre-tread applications ranged in iodine number from 80 (HAF N330) to 140 (SAF N110). The most widely used tread blacks were N330 (HAF), N220 (ISAF), and N347 (HAF-HS). The superior treadwear theoretically associated with the highest-area tread black available (N110 SAF) was generally not commercially exploited as a result of problems associated with high area: i.e. difficult processing, groove cracking in cross-ply tyres and high cost. Therefore, the chief industry demand was for blacks which would give better treadwear at the same surface area or iodine number.

This demand was first met in the early 1970s by the introduction of 'new technology' or 'improved' blacks.[11,41-43] Since different manufacturers employ different furnaces and other variations in technology, the blacks of a given grade are not completely equivalent, and may owe their improved treadwear/surface area relation to different combinations of manufacturing techniques and measured properties.

One route[11,41,42] employed furnaces which give better control of feedstock atomisation and closer localisation of the different processes taking place in the reactor (as compared with the reactors used to make the earlier 'conventional' grades). This leads to a narrower distribution of aggregate sizes, as shown by centrifugal sedimentation[11,44] and by image analysis of electron micrographs.[44] Another result of the increased sharpness of these furnace processes is that the aggregates stabilise in relatively more open configuration, rather than filling up with more carbon as in the clustered morphology of the older ('conventional') blacks. This more open configuration permits attainment of a given bulkiness or DBPA with fewer particles per aggregate (N_p) and thus smaller aggregate mass at a given surface area (Section 1.2).[11,44] Because of the smaller aggregate mass, and to a lesser extent because of the narrower distribution of aggregate sizes, the new technology blacks have a higher tinting strength (Section 1.3) than conventional blacks of the same surface area and DBPA.

The new technology blacks also differ from conventional blacks in their surface properties, being characterised by higher surface activity (higher bound rubber and higher moisture adsorption) and a lower ratio of iodine number to external surface area ('t' or CTAB).[11] The latter factor accounts for *part* of the improvement in treadwear at a given iodine number, but a very significant improvement (up to 7%) is found even at the same external surface area.[11] This improvement may be ascribed to the following factors:

1. higher surface activity;
2. narrower distribution of particle size and virtual absence of very large particles[11] which are believed to act as weak spots in abrasion or tensile failure;[7,45]
3. more open aggregates.

Because of the larger number of more open aggregates, a stronger inter-aggregate network is set up,[11,46] leading to slightly higher hysteresis,[11,41] which appears to contribute to the improved treadwear. The difference in hysteresis is definitely noticeable if the blacks are compared at equal iodine number but is only marginal at equal external area.[47]

'Improved' blacks[43] have also been reported to have fewer large, clustered aggregates than conventional blacks.[48] At equal nitrogen area these blacks give the same rebound (or hysteresis) as conventional blacks.[49] 'In general, improved blacks show less breakdown in rubber and greater bound rubber development';[43] however, the relative magnitude of these factors depends on the grade. The improved treadwear is attributed primarily[43,49] to increased polymer/filler interaction as revealed by higher bound rubber and higher torque development in the Banbury mixing cycle. The increased torque, in turn, is believed to lead to a more effective dispersion of the black. Hess and Chirico[50] have also demonstrated by image anaysis that the 'improved' blacks retain more bound polymer after dispersing the gel.

3. PRODUCTION AND QUALITY CONTROL

3.1. Lampblack Process

The Chinese are believed to have developed the first carbon-black process for pigmentation applications over 5000 years ago. They simply burned vegetable oil in small lamps with tile covers on which the carbon black accumulated. From this ancient practice has evolved the lampblack process. This process features the partial combustion of petroleum and coal-based feedstocks in large open pans. Lampblacks are characterised by high structure and low surface area.[51] Small quantities are still produced today.

3.2. Channel, Thermal, Acetylene and Gas-furnace Processes

The channel process for carbon-black production has had a long and successful history, beginning in 1872. World-wide production of channel black has fallen sharply in recent years because of rising natural-gas prices, smoke pollution, low yield, unsatisfactory performance in synthetic-rubber tread compounds, and the rapid development of the oil-furnace process. The name channel black comes from the use of steel channel irons. Carbon black is impinged on the flat side of the channels in the proximity of thousands of small natural-gas flames. The carbon black deposited on these channels is scraped off and collected. Channel black is characterised by high oxygen content and low structure.

Both the thermal and acetylene black processes are based on the high-temperature decomposition of hydrocarbons in the absence of air or flame. Thermal black is produced from natural gas in an endothermic cracking reaction requiring a large heat-energy input. Heat is provided by firing the thermal generator with a stoichiometric ratio of air and gas. Heat is absorbed by the refractory lining and released to crack the natural gas during the next 'make' cycle. Two thermal generators are used alternately. When a producing generator becomes too cool for carbon-black formation it is placed back on a heat cycle. While basically a batch process, the alternating cycles of approximately 5 min duration, and a carbon yield of 40–50%, provide an essentially continuous flow of product. Thermal black is noted for very low structure and low surface area or very large particle size. Most of the world's thermal production is in North America where natural gas has been plentiful. Production of thermal black has dropped in recent years with rising gas prices and decreasing availability.

Acetylene black is made by the continuous exothermic decomposition of acetylene gas in water-cooled, refractory-lined metal retorts. A unique form of carbon black is produced at the high reaction temperature and long residence time of this process. The product has an unusually high degree of structure and gives high electrical and thermal conductivity in rubber compounds. A major use of acetylene black is in dry-cell batteries.

In 1922 a continuous carbon-black process utilising the partial combustion of natural gas was developed. The basic concepts used in the combustion, collection, pelletisation and handling of gas-furnace blacks were later applied to the oil-furnace process. A major disadvantage of the gas-furnace process was the inability to produce highly reinforcing carbon blacks required for long tyre-tread life. Channel blacks were acceptable for tyre treads as long as natural rubber was in use. They did not perform well in non-crystallising synthetic rubbers with their greater need for filler reinforcement. Therefore, the development of synthetic polymers reduced the acceptability of both channel and gas-furnace blacks. Additionally, the rapid rise in natural-gas prices adversely affected the production costs of both channel and gas-furnace blacks. These factors set the stage for the development of a new process that could produce more reinforcing carbon blacks from a raw material which had better price stability and was readily transportable.

3.3. The Oil-furnace Process

A patent for producing carbon black from oil was issued to W. H. Frost in 1922. The first oil-furnace blacks were commercially produced in 1943, after a sharp rise in natural-gas prices. The oil-furnace process today is the most significant and most versatile means of producing carbon black.

A schematic of a modern oil-furnace unit utilising the wet pelletising process is shown in Fig. 3. After preheating, the liquid hydrocarbon feedstock is injected into one or more reactors. Here it is atomised and mixed with preheated air and an auxiliary fuel and heat source, normally natural gas. The reaction is quickly quenched with a water spray (see Section 1.1).

The product is then separated from the combustion gases by a series of bag filters and cyclones. From here the fluffy black is carried to a pelletiser where water and if necessary a small amount of pellet binder is added. A typical pelletiser is a pin-type mixer consisting of many metal pins arranged helically around a central shaft. Rotation of the shaft creates a mixing action that forms dense, spherical carbon-black pellets.

The wet pellets pass out of the pelletiser and are then fed into a rotary dryer to remove the pelletising water. After drying, the carbon-black pellets are passed over a magnetic separator to remove magnetic material that may have been picked up within the system. The pelleted black then passes to a classifier to remove over-size pellets and finally the black is conveyed by belts, screw conveyors or bucket elevators to stor-

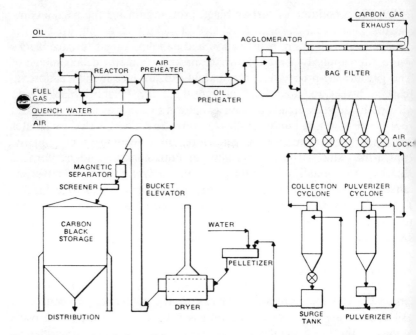

Fig. 3. Schematic of the oil-furnace process. (From *Hydrocarbon Processing*, September 1977, pp. 165–173).

age silos. From here it is either packed into bags or bins or loaded into bulk hopper cars or trucks for shipment.

3.4. Quality Control

An effective quality-control programme is a major factor in maintaining high and consistent product performance. Major control areas include raw-material evaluation, process control and product analysis.

A highly aromatic, high-viscosity hydrocarbon is the primary feedstock for producing carbon black by the oil-furnace process. The purity of the feedstock is important, especially with respect to sulphur content, alkali-metal content and total ash content. Viscosity is measured to determine flow properties. Other physical and chemical properties used to determine quality include specific gravity, asphaltine content, and US Bureau of Mines 'correlation index'.

The next phase in an effective quality-control programme calls for close control of all stages of the oil-furnace process. The flow rates of oil feedstock, natural gas, combustion air and flame additive, if used for control of structure, need accurate metering and control. In order to obtain the desired pellet properties, the flow rates of fluffy black, water, and pelletising additive if used are critical and also require close control. The pellets must be dense and firm, for efficient storage, shipping and handling in bulk, but still be soft enough to break up readily and disperse in rubber compounds. After wet-process pelletising, the product passes through a rotary dryer where temperature control is important. Too low a temperature will give wet pellets, while a high temperature can influence the oxygen content and surface activity of the black and change rubber properties.

Product analysis for process-control purposes is normally done on fluffy or unpelleted black samples taken immediately downstream from the reactor. Typical tests here include iodine adsorption to measure surface area, DBP absorption for structure control, and solvent discoloration. Frequent samples are normally taken at this point in order that reactor inputs can be adjusted as needed. More extensive testing is performed on dried pellets taken at the dryer exit. In addition to the above tests, dryer samples may be evaluated for tinting strength, ash content, foreign material, moisture content, dust content, mass pellet strength, individual pellet crush strength, density, pH, sulphur content and compound rubber properties.

3.5. Quality Assurance

The carbon-black industry has made great strides in recent years in upgrading quality-assurance programmes. This has resulted directly from demands from the rubber industry for better and more uniform product performance and in-plant processing. While the procedures will vary from one producer to another, the ultimate goal is the same—to assure the user that product shipments conform to the user's specifications. Test data generated for quality control may be augmented with results on samples taken at the time of shipment. Multiple shipment samples normally are checked for pellet quality, moisture content, iodine adsorption and DBP absorption to assure product uniformity. A typical quality-assurance programme may include procedures for handling customer questions and complaints on quality. Generally a technical representative will coordinate such discussions between the producer and user. More open communication in the areas of material specifications, rubber product requirements and rubber plant processing has had a favourable impact on the upgrading of carbon-black quality.

4. SELECTION OF CARBON BLACK FOR COMPOUND PERFORMANCE

4.1. Introduction

The selection of a carbon-black grade for a particular product application is governed by the two main characteristics of carbon black, surface area and structure.

TABLE 1

EFFECTS OF CHANGES IN SURFACE AREA AND STRUCTURE ON RUBBER PROPERTIES

	Increasing surface area	*Increasing structure*
Processing properties		
Loading capacity	Decreases	Decreases
Incorporation time	Increases	Increases
Oil-extension potential	Little	Increases
Dispersability	Decreases	Increases
Mill bagging	Increases	Increases
Viscosity	Increases	Increases
Scorch time	Decreases	Decreases
Extrusion shrinkage	Decreases	Decreases
Extrusion smoothness	Increases	Increases
Extrusion rate	Decreases	Little
Vulcanisate properties		
Rate of cure	Decreases	Little
Tensile strength	Increases	Decreases
Modulus	Increases to maximum then decreases	Increases
Hardness	Increases	Increases
Elongation	Decreases to minimum then increases	Decreases
Abrasion resistance	Increases	Increases
Tear resistance	Increases	Little
Cut-growth resistance	Increases	Decreases
Flex resistance	Increases	Decreases
Resilience	Decreases	Little
Heat build-up	Increases	Increases slightly
Compression set	Little	Little
Electrical conductivity	Increases	Little

In general, the higher the surface area of the carbon black then the higher is the reinforcement in terms of tensile strength, tear strength, abrasion and fatigue resistance. Hysteresis losses are also higher giving rise to increased temperature generation during mixing and in products with dynamic applications.

Structure is more important in terms of compound processability, giving better carbon-black dispersion and improved compound extrusion characteristics. Cured modulus and hardness increase with structure and this can lead to inferior cracking and fatigue performance. However, treadwear under severe conditions is improved by higher structure.

A summary of the effects of these two parameters is given in Table 1. The relative importance of these uncured and cured properties is very dependent on the product application.

4.2. Tyre Applications

Tyre production still consumes approximately 75% of all carbon black produced and therefore exerts the most influence on the grades produced and in particular, those at the reinforcing end of the carbon-black spectrum. It is therefore appropriate that we first examine this application considering the various tyre components in turn.

4.2.1. Car-tread Compounds

Car-tread compounds have the requirement of high abrasion resistance, good wet grip and generally low compound cost. Up to a few years ago, car treads were based on N330 (HAF) or N220 (ISAF) but with the advent of new technology blacks and their improved abrasion per unit cost the majority of this market has switched to use of N339 (Improved HAF) or N375 (Improved HAF). Both of these blacks have an abrasion resistance approximately equal to ISAF but priced equal to N330 (HAF), and have been used with increased oil loadings giving improved wet-grip performance and lower compound cost.

N234 has recently been added to the range of new technology blacks. This grade has a reinforcement/abrasion level approximately 10% higher than N220 (ISAF) but at a cost equal to N220 (ISAF). It has found application for premium tread compounds where improved abrasion is still being requested but by using it at higher loadings of black and oil it offers the possibility of improved wet grip while retaining normal treadwear performance.

4.2.2. Truck-tread Compounds

Truck treads in addition to good wear must have satisfactory temperature generation, flaking, cutting and chipping resistance. These last three parameters are becoming particularly important for European truck-tyre usage where off-the-road performance is becoming an additional requirement.

Flaking, cutting and chipping are adversely affected by high carbon-black structure and tread compounds historically were based on the normal-structure blacks N330 (HAF), N220 (ISAF) and some N110 (SAF). The majority of new technology carbon blacks have high structure, and special compounding techniques need to be employed to take advantage of their good abrasion/cost ratio. The compound modulus must be dropped to acceptable levels and this can be done by reducing the crosslink density by modifying the cure system, by substituting up to 15 parts of the carbon black by a precipitated silica, or by introducing SBR in place of natural rubber where the increased temperature generation can be accepted. By use of these techniques N339, N375 or N234 can successfully be used.

For pure off-the-road performance as in earthmover and loader dozer applications, normal abrasion resistance in the tread compound is less important and rubber loss by flaking, cutting and chipping becomes paramount. Typical carbon-black usage for this has been N110, N231 or blends of N110/N326. However, increasingly SBR is being used for slow-moving applications where temperature generation permits, and blends of N110 and silica in 100% natural rubber are being used for the higher-speed, general-purpose application.

The possibility that new technology blacks may be used in pure off-the-road application will depend either on modifications to existing structure levels, or to the economic acceptance of lower than normal black loading levels.

4.2.3. Bonding Compounds

Although not subjected to direct abrasion, the bonding compounds used in breaker constructions have a requirement for high reinforcement. Initially, bonding compounds for both textile and brass were based on 100% natural rubber and used channel-black reinforcement. The channel black produced slow-curing characteristics which appear to give more latitude for the compound crosslink reaction and the brass/sulphur reaction to be coincident. When channel black was eliminated from the carbon-black range due to cost and pollution problems, its initial replacement was by an oxidised, low-structure HAF (S315) which had similar slow-cure characteristics. Eventually the S315 was withdrawn for production and cost reasons and has now generally been replaced by normal cure rate, low-structure HAF (N326).

The slow-cure characteristics are now obtained by suitable choice of a slow-curing accelerator or by part replacement of the carbon black by silica/resorcinol/hexamine systems. Such compounds appear to be meeting design requirements and there is no indication of significant changes being made in Europe in the near future. However, in the USA there appears to be a lower performance requirement for breakers and N351 has found some application.

4.2.4. Sidewall Compounds

Cracking and fatigue performance are most important for sidewall compounds but some degree of abrasion resistance is also necessary. Typical carbon-black choice is either N330 or N550 (FEF) but there is no reason why new technology blacks should not be introduced providing a low loading is used or other compounding techniques adopted in order to keep a satisfactory modulus level.

4.2.5. Casing Compounds

Tyre-casing compounds have traditionally been based on use of N660 and this is still satisfactory for remaining cross-ply tyres and car-radial casings. However, there is increasing interest in part or total replacement of carbon blacks by cheaper mineral fillers such as clay as the price of carbon black increases. Higher-performance tyres such as aero and mono-ply steel truck casings often require the additional reinforcement and fatigue resistance offered by N330 or N326.

4.2.6. Inner Liner

The role of the inner liner is air retention and carbon black has little effect on permeability. However, it can have significant effect on dimensional stability and for this the high-structure blacks N550, N683 (APF) and N650 (GPF–HS) would be preferable.

4.2.7. Run-flat Tyres

Run-flat tyres are now a commercial reality although there has been as yet little market penetration. There is no reason to expect that this type of tyre will require any special type of carbon black in the tread or breaker assemblies but they could eventually have significant influence on the pattern of semi-reinforcing blacks.

Run-flat tyres obtain much of their run-flat performance from the use of thick sidewall sections capable of supporting the tyre load. These compounds will require high modulus in order to reduce the tyre deflection, moderate reinforcement to survive the fatigue requirement and obviously low temperature-generation characteristics. This is undoubtedly a compromise situation and high-structure blacks of the N650, N683, N351 or N550 appear likely candidates. If this type of tyre becomes established then the long-term viability of N660 black would be in doubt.

4.2.8. Low Rolling Resistance
The oil crisis has already prompted the United States to introduce legislation on energy saving, including tyres. Much of the savings are being achieved by tyre-design changes but reduced energy loss compounds are probably also required. With current radial-tyre designs, little reduction can be achieved in the casing, sidewall or breaker compounds and attention is therefore being concentrated on tread compounds. Reduced energy loss could easily be achieved by moving to a less-reinforcing carbon black but significant loss in treadwear does not appear acceptable. A new range of carbon blacks is currently of interest which has marginally better resilience relative to treadwear. These new blacks have so far generated very little interest in Europe due to the high wet-grip requirement but with further development and more significant resilience advantage but with increasing emphasis on energy savings through reduced rolling resistance they could become of interest.

4.3. Non-tyre Applications

Conveyor-belt compounds have a similar requirement to tyre-tread compounds in terms of reinforcement. Carbon-black usage has therefore followed a similar pattern and the new technology grades are well established. The latest type, N234, is already of interest in that the higher reinforcement allows SBR compounds to be used for high-grade covers.

In most of the non-tyre applications, high abrasion resistance is less important and is replaced by modulus and hardness development, dynamic properties, processability and compound cost. Carbon-black usage varies from the 300 series down to the very coarse particle size blacks and including lampblack and thermal blacks.

The coarse, low-structure thermal blacks find applications where high loadings are important as a polymer diluent for either performance or cost reasons. Product applications in this group are oil seals, door seals, mat, etc.

Lampblack, due to its coarse particle size and high structure, is used for modulus and hardness development combined with high resilience in compounds such as engine mountings and tank-tread compounds.

N762 is widely used in hose applications for its good processing characteristics, and the high-structure semi-reinforcing blacks such as N683 and N550 find application in extruded products where high black and oil loadings are required for cost reasons.

There are no significant improvements expected in the semi-reinforcing blacks but rather a contraction of the range due to cost or environmental reasons. Fine thermal black has already virtually ceased production and similar pressures must exit on medium thermal and lampblack.

As a consequence of the loss or potential loss of the coarser grades, many compounders have been substituting their use by normal furnace grades. Medium thermal can be replaced by either SRF or GPF in combination with white fillers. Lampblack can be satisfactorily replaced by N683, N650 or N550 in many applications.

4.4. Energy Loss/Temperature Generation

In the preceding section, reference has been made to the need for controlling energy loss and temperature generation for various applications. Pendulum resilience or flexometer heat build-up measurements have for many years been used as the main criteria for predicting heat-generation performance. However, dynamic moduli, together with a better knowledge of the mode of operation of any product, can now offer improved compound design for this particular feature.

In the Kelvin model (Fig. 4) the elastic modulus E' is represented by a spring and the loss modulus E'' by a dashpot.

Resilience for a sinusoidal oscillation can be shown to be

$$R = \exp(-\pi E''/E')$$

Or if E''/E' is small then

$$R = 1 - \pi E''/E'$$

$$(1 - R) = \pi E''/E'$$

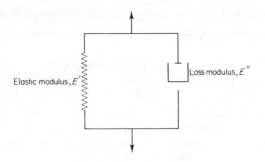

Fig. 4. Kelvin model.

The energy loss in any system is $(1 - R) \times$ energy input. Three types of energy input condition can be considered and it is relatively simple to show the following dependence of energy loss on the dynamic moduli:

Energy loss at constant load is proportional to $E''/(E')^2$.
Energy loss at constant deflection is proportional to E''.
Energy loss at constant energy is proportional to E''/E'.

From this it can be seen that it is extremely important to determine the mode of operation of a component before deciding the requirements of compound dynamic moduli. For example, a truck-tread compound operates largely under constant load condition and its temperature generation is extremely dependent on the ratio $E''/(E')^2$ whereas the sidewall components of the tyre operate under constant deflection and are dependent on E'' only.[52] The loss modulus is therefore important in both components but the elastic modulus is only of significance in the tread compound.

Dynamic moduli vary with temperature, frequency and amplitude (see Fig. 5 for effect of amplitude) and it is therefore extremely important to carry out dynamic-moduli measurements under the condition of service operation that the compound will undergo.

Carbon black significantly affects both moduli; increasing surface area increases the elastic and loss moduli, and increasing structure increases the elastic modulus. As reinforcement is invariably affected by particle size, energy loss/temperature generation and reinforcement are often a matter of compromise.

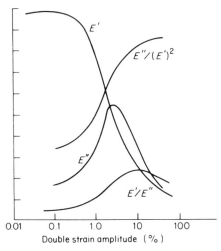

FIG. 5. Strain amplitude dependence of dynamic parameters.

4.5. Rationalisation

In any presentation on carbon-black-grade selection it would be wrong not to consider grade rationalisation and the need to keep the number of grades down to a minimum. Carbon black is produced on a continuous process which operates to the best quality and economics when run for long periods at a steady state. The number of grades is extremely important from the bulk storage consideration as additional storage capacity requires additional capital expenditure. It is important to understand that modified pellet properties constitute a separate grade of carbon black equally as much as a modification to the analytical properties.

With the introduction of any new type of carbon black it is essential that both the producer and user have identified sufficient gain in performance that the new grade can develop into a major market grade otherwise it will have no long-term viability.

With the introduction of any major new grade, most producers, due to their limited storage capacity, must also look to the elimination of one of their major existing grades. For this reason, it is essential that any compounder must keep well aware of any market trends, designing new compounds on viable grades and modifying compounds away from grades which are in danger of elimination.

5. FUTURE PROSPECTS

5.1. General
Carbon black will continue to be an essential raw material for the rubber industry into the indefinite future. Modern transportation systems depend on tyres, and tyres depend on carbon black for reinforcement. While carbon blacks have been partially replaced by other fillers in certain applications, there are no known substitute materials which can satisfactorily and economically replace carbon black in tyre-tread compounds. Growing demands for improved tyre-tread life, greater road holding and better fuel economy will increasingly require the versatility in rubber performance afforded by carbon blacks.

5.2. Production
The oil-furnace process will continue to be the primary means of producing carbon black. Further improvements are expected in areas of greater plant efficiency, energy conservation and reduced air pollutants from stack emissions. Automated control systems are expected to become more sophisticated as efforts to maintain uniform quality levels continue. Continuing technological advances are expected to further broaden the versatility of the oil-furnace process and lead to new products for tomorrow's needs within the rubber industry.

5.3. Application Areas
The general market distribution for carbon black from the year 1910 is shown in Fig. 6. It can be seen that since around 1960 the market share of rubber and non-rubber uses has been stable. This is also generally true of the carbon black split between tyre and non-tyre rubber applications. In future years it seems likely that with new developments, both non-rubber uses and non-tyre rubber uses will grow somewhat faster than the usage in passenger car tyres. Plastics, for example, are becoming a significant application for carbon black. Some plastics applications require grades of higher surface area, higher surface oxygen content, or higher electrical conductivity than those used in the rubber industry. Rubber-extended paving asphalt is under test. If adopted, this could consume large quantities of carbon black. Smaller cars in the USA, rising petrol prices, and longer-wearing tyre designs and compounds world-wide seem likely to moderate the consumption growth rate of carbon blacks in car tyres. Usage in truck and off-the-road service should continue to grow in line with general economic conditions.

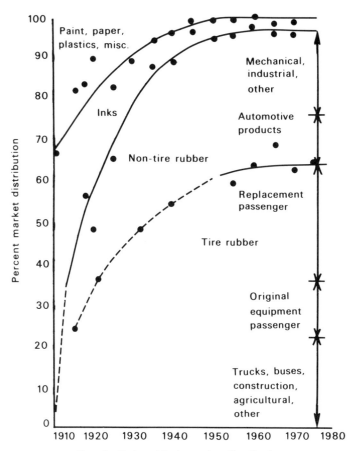

FIG. 6. Carbon-black market distribution.

5.4. Grade Distribution

The USA production of carbon blacks by process and by general grade type is shown in Fig. 7. The carbon-black grade distribution has changed as process innovations and new consumer requirements gave rise to new improved qualities. The new technology blacks have been most popular in the N300 range. This is reflected in the surge of the HAF grades at the expense of ISAF in the early 1970s. By 1977 this trend had begun to reverse itself as certain tyre producers have moved to N234 black to meet treadwear requirements in critical tyre designs. N200 grades are expected to maintain a significant market share as N234 and possible longer-wearing grades are adopted where needed.

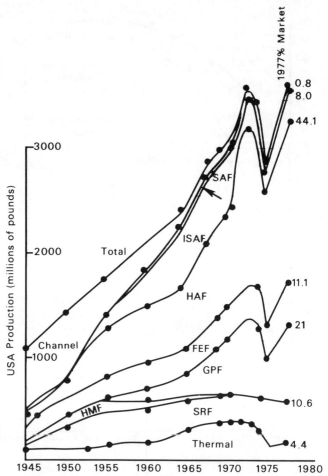

FIG. 7. Carbon-black production in the USA by process and grades.

N600 grades have also increased significantly since 1965. N660 has largely replaced the gas-furnace HMF black, N601. Also, GPF–HS, N650, has replaced FEF black in several important compounding areas. Further increases in the GPF types are forecast, with this growth coming at the expense of thermal and SRF types as well as FEF.

5.5. New Products
Work is already well advanced to produce a series of carbon blacks having improved resilience without loss in treadwear. This development

is expected to lead to a new series of commercial carbon blacks for the tyre industry. Research is continuing for the development of longer-wearing carbon blacks. A new black giving 5–10% longer treadwear than N234 is very possible. New grades are also probable for improved tear and chipping resistance in off-the-road tyre service and truck tyres.

For the non-tyre market future products are anticipated in the areas of highly conducting carbon blacks. Recently a black has been introduced which has outstanding electrical properties enabling high conductivity to be obtained by addition of only a few parts per hundred of polymer. The main use so far has been in plastics but this black could be particularly beneficial in antistatic fireproof compounds.

Also in the non-tyre field further reductions in the thermal and lampblack availability will increase demand for oil-furnace-based alternatives.

5.6. Reclaimed Carbon Black

Reclaim of oil, carbon black and metal from scrap tyres by pyrolysis[53] is currently under investigation and could become a commercial reality. The black compounded in the scrap tyre is a blend of reinforcing and semi-reinforcing type dependent on the original tyre manufacturer, the type of tyre, and the extent of treadwear. The final product from pyrolysis is therefore not uniform in its properties and additionally contains carbonisation products from the rubber and a high ash content. The carbonisation products tend to cement carbon-black aggregates together, hampering good dispersability. At the current state of the art, the recovered black does not appear to be a practical reinforcing filler but considerable effort is being expended on improving the recovery process and superior properties have been claimed.

5.7. Carbon Black Handling—Pellet Properties

Due to the requirements of increased efficiency, lower cost and improved working environment many small and intermediate-size companies are installing bulk-handling systems. These can be mechanical or pneumatic in operation and they vary in their attrition of the carbon-black pellets.

In order to minimise the amount of breakdown, customers with 'high attrition' systems require a very stable pellet but all users require a pellet which will disperse well during the Banbury mixing cycle. A compromise situation will be required.

REFERENCES

1. (a) SWEITZER, C. W. and HELLER, G. L. *Rubber World*, **134**, 1956, 855; (b) HOMANN, K. H. *Angew. Chem. Intern. Edit.*, **7**, 1968, 414; (c) LAHAYE, J., PRADO, G. and DONNET, J. B. *Carbon*, **12**, 1974, 27.
2. FRIAUF, G. F. and THORLEY, B. US 3010794; US 3010795, publ. 28 November 1961.
3. (a) ERGUN, S. *Carbon*, **6**, 1968, 141; (b) HARLING, D. F. and HECKMAN, F. A. *Mat. Plast. Elast.*, **35**, 1969, 80.
4. SCHAEFFER, W. D. and SMITH W. R. *Ind. Eng. Chem.*, **47**, 1955, 1286.
5. RIVIN, D. *Rubber Chem. Technol.*, **44**, 1971, 307.
6. HAWS, J. R., COOPER, W. T. and ROSS, E. F. *Rubber Chem. Technol.*, **50**, 1977, 211.
7. McDONALD, G. C. and HESS, W. M. *Rubber Chem. Technol.*, **50**, 1977, 842.
8. SMITH, N. L. *Rubber Age*, *Rubber World*, May 1970 (advertisement).
9. (a) ASTM Test D 1765-77; (b) ASTM Test D 2516-76.
10. MEDALIA, A. I. *J. Colloid Interface Sci.*, **32**, 1970, 115.
11. MEDALIA, A. I., DANNENBERG, E. M., HECKMAN, F. A. and COTTEN, G. R. *Rubber Chem. Technol.*, **46**, 1973, 1239.
12. MEDALIA, A. I. and EATON, E. R. *Kautsch. Gummi Kunstst.*, **20**, 1967, 61.
13. BRUNAUER, S., EMMETT, P. H., and TELLER, E. *J. Am. Chem. Soc.*, **60**, 1938, 309.
14. ASTM Test D 3037-77.
15. ASTM Test D 1510-76.
16. JANZEN, J. and KRAUS, G. *Rubber Chem. Technol.*, **44**, 1971, 1287.
17. ABOYTES, P. and VOET, A. *Rubber Chem. Technol.*, **43**, 1970, 464.
18. DE BOER, J. H., LINSEN, B. G., VAN DER PLAS, TH. and ZONDERVAN, G. J. *J Catalysis*, **4**, 1965, 649.
19. SMITH, W. R., and KASTEN, G. A. *Rubber Chem. Technol.*, **43**, 1970, 960.
20. (a) SALEEB, F. Z. and KITCHENER, J. A. *J. Chem. Soc.*, **1965**, 911; (b) ABRAM J. C. and BENNETT, M. C. *J. Colloid Interface Sci.*, **27**, 1968, 1.
21. ASTM Test D 2414-76.
22. (a) VOET, A. and WHITTEN, W. N. Jr. *Rubber World*, **146**, 1962, 77; (b MEDALIA, A. I. *Rubber Age*, **93**, 1963, 580.
23. DOLLINGER, R. E., KALLENBERGER, R. H. and STUDEBAKER, M. L. *Rubber Chem. Technol.*, **40**, 1967, 1311; (b) ASTM Test D 3493-76.
24. MEDALIA, A. I. and RICHARDS, L. W. *J. Colloid Interface Sci.*, **40**, 1972, 233
25. GESSLER, A. M. *Rubber Age*, **101**, 1969, 54.
26. MEDALIA, A. I. and RIVIN, D. In *Characterization of Powder Surfaces*, G. D Parfitt and K. S. W. Sing (Eds.), Academic Press, London–New York–Sa Francisco, 1976, Chapter 7.
27. (a) EINSTEIN, A. *Ann. Physik*, **19**, 1906, 289; (b) *ibid*, **34**, 1911, 591.
28. (a) SMALLWOOD, H. M. *J. Appl. Phys.*, **15**, 1944, 758; (b) GUTH, E. *J. Appl Phys.*, **16**, 1945, 20.
29. RUTGERS, R. *Rheol. Acta*, **2**, 1962, 202, 305.
30. (a) GUTH, E., and GOLD, O. *Phys. Rev.*, **53**, 1938, 322; (b) GUTH, E. *Proc. 5t. Intern. Congr. Appl. Mechanics*, Cambridge, 1938, p. 448.
31. KRAUS, G. *J. Polymer Sci.*, **B8**, 1970, 601.

32. MEDALIA, A. I. *Rubber Chem. Technol.*, **45**, 1972, 1171.
33. MEDALIA, A. I. *Rubber Chem. Technol.*, **46**, 1973, 877.
34. (a) PARKINSON, D. *Trans. Inst. Rubber Ind.*, **16**, 1940, 87; (b) *ibid*, **19**, 1943, 131; (c) *ibid*, **21**, 1945, 7.
35. MEDALIA, A. I. *Rubber Chem. Technol.*, **51**, 1978, 437.
36. PAYNE, A. R. In *Reinforcement of Elastomers*, G. Kraus (Ed.), Interscience Publishers, New York, 1965, Chapter 3.
37. LAKE, G. J. and THOMAS, A. G. *Proc. Royal Soc.*, **A300**, 1967, 108.
38. BUECHE, F. *Physical Properties of Polymers*, John Wiley & Sons, New York, 1962.
39. (a) DANNENBERG, E. M. *Trans. Inst. Rubber Ind.*, **42**, 1966, T26; (b) RIGBI, Z. *Koll.-Z.*, **224**, 1968, 46.
40. KRAUS, G. *Rubber Chem. Technol.*, **44**, 1971, 199.
41. DANNENBERG, E. M. *J. Inst. Rubber Ind.*, **5**, 1971, 190.
42. JORDAN, M. E., BURBINE, W. G. and WILLIAMS, F. R. US 3 725 103, publ. 3 April 1973; *idem*, US 3 799 788, publ. 26 March 1974.
43. HESS, W. M., CHIRICO, V. E. and BURGESS, K. A. Paper presented at the *Intern. Rubber Conf., Prague, Czechoslovakia*, September 1973.
44. REDMAN, E., HECKMAN, F. A. and CONNOLLY, J. E. Paper presented at a meeting of the *ACS Div. Rubber Chem., Chicago*, May 1977; abstract in *Rubber Chem. Technol.*, **50**, 1977, 1000.
45. STACY, C. J., JOHNSON, P. H. and KRAUS, G. *Rubber Chem. Technol.*, **48**, 1975, 538.
46. MEDALIA, A. I. *Rubber World*, **168**, 1973, 49.
47. CARUTHERS, J. M., COHEN, R. E. and MEDALIA, A. I. *Rubber Chem. Technol.*, **49**, 1976, 1076.
48. HESS, W. M., McDONALD, G. C. and URBAN, E. *Rubber Chem. Technol.*, **46**, 1973, 204.
49. DIZON, E. S., MICEK, E. J. and SCOTT, C. E. *J. Elastomers Plastics*, **8**, 1976, 414.
50. HESS, W. M. and CHIRICO, V. E. *Proc. Inst. Rubber Ind., 1st Eur. Conf., Brussels, Belgium*, 1975.
51. DANNENBERG, E. M. In *Vanderbilt Rubber Handbook*, 12th ed., R. T. Vanderbilt Co., New York, 1977.
52. COLLINS, J. M., JACKSON, W. L. and OUBRIDGE, P. S. *Rubber Chem. Technol.*, **38**, 1965, 2.
53. CRANE, G., ELEFRITZ, R. A., KAY, E. L. and LAMON, J. R. *Rubber Chem. Technol.*, **51**, 1978, 577.

Chapter 5

SILANE-TREATED MINERAL FILLERS IN RUBBERS

EDWIN P. PLUEDDEMANN
Dow Corning Corporation, Midland, Michigan, USA

and

BRYAN THOMAS
Dow Corning Corporation, Barry, UK

SUMMARY

Low-cost mineral fillers have been upgraded with organofunctional silanes to obtain creditable performance as reinforcing fillers in rubber.

Silane-treated fillers generally have improved dispersion in the compound, and may modify the rate of cure, but the primary advantage of the treatment is related to interfacial adhesion between polymer and filler. Initial tests on adhesion suggest that properties of mineral-filled rubbers might be improved even more if greater adhesion were obtained at the interface. This improved adhesion requires a silanol-modified resinous phase at the filler surface. Some degree of resinous nature (reduced mobility) is probably imposed on rubber molecules bound to high-energy surfaces, but it is possible that total performance could be improved still more by selecting a silane, or a silane-modified resin precursor that will provide a high degree of adhesion between rubber and mineral as demonstrated by microscope-slide adhesion tests.

1. INTRODUCTION

Although mineral fillers have been used in organic rubbers for many years, their total application has been small compared to carbon blacks.

Low-cost minerals are not such effective reinforcing fillers as carbon black, and high-surface silicas that are good reinforcing fillers are more costly than carbon black. Only in silicone rubbers have costly high-surface silicas become the standard reinforcement.

During the last few years the carbon-black industry has been severely affected by the energy situation, and the end is not in sight for energy-related products. For example, thermal black, a natural-gas product, has become a premium commodity, and is no longer the cheap extender pigment of a few years back. Its use is now principally in products where its unique properties are essential. Oil-furnace grades of carbon black have increased in price but not to the point of satisfactory profitability so further increases are anticipated.

Precipitated silica pigments, although not petrochemical-based like carbon blacks, are energy-sensitive because of the power requirements necessary for reacting, precipitating and drying. Clays and calcium carbonates, being much less energy-dependent, are not as susceptible to sharp price increases. Energy requirements for the various fillers are listed in Table 1.[1]

TABLE 1

	Energy requirements, BTUs/lb	1978 Price, $/lb
Carbon black—soft	37 800	0·25
Carbon black—hard	50 000	0·28
Carbon black—medium thermal	58 300	0·34
Calcined clay	5 800	0·20
Silane-treated clay	1 700	0·20
Precipitated silica	13 500	0·44

Clays and calcium carbonates are among the most abundant and lowest-cost minerals on earth and need only to be mined and beneficiated to make them suitable for use by the rubber industry.

It is anticipated that these materials will play a much larger role in future rubber technology. Calcium carbonate is not a reinforcing filler and is used only as an extender as in blends with soft carbon black to replace thermal blacks.

Fine particle silica has chemical reactivity with free radicals comparable with that of carbon black. Formation of bound rubber is comparable when these two fillers are milled in SBR or butyl rubber. It appear that a fine particle silica undergoes chemical filler/polymer interaction similar to HAF carbon blacks in rubber formulations.[2]

Much effort has been expended in up-grading low-cost mineral fillers by treating them with reactive silanes in order to obtain a cost-effective replacement for carbon black. Although much progress has been made in formulating rubbers with silane-treated mineral fillers it is believed that significant improvements can still be made in present systems. As the pressure continues on energy and carbon-black costs, there will be greater incentive to develop silane-modified mineral fillers as replacements.

2. CONCEPTS OF RUBBER REINFORCEMENT

2.1. Rubber/Filler Bonds

It is generally agreed that stong links exist between rubber chains and reinforcing-filler particles. By participating in crosslinking, the fillers become coated with a layer of bound rubber of higher modulus than the more remote rubber matrix.[3]

The effect of a strong chemical bond between matrix and filler on tensile properties of filled urethane rubber was demonstrated by comparing glass microbeads (poor adhesion) with a ground epoxy resin (good adhesion) as filler.[4] Tensile failure in the rubber was preceded by vacuole formation in the matrix. Thin films at the filler surface then ruptured and continued to fail through a peeling mechanism. A soft matrix at the filler surface facilitated vacuole formation at lower stresses resulting in poor strength. Fillers surrounded by a higher-modulus layer of polymer slowed vacuole formation and resisted initial dewetting and peeling. Fillers that do not provide extra crosslinking and do not bond to the rubber are non-reinforcing fillers or extenders.

2.2. Surface Treatment

Surface modification of mineral fillers with 'coupling agents' in rubber has generally been directed towards improved mechanical properties of the composite as related to improved adhesion across the interface.

Although adhesion is central to any 'coupling' mechanism, it is recognised that many factors are involved in the total performance of a composite system. The interface, or interphase region, between polymer and filler involves a complex interplay of physical and chemical factors related to composite performance as indicated in Fig. 1.

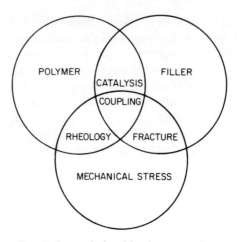

FIG. 1. Interrelationships in composites.

It is recognised that the total coupling mechanism involves all of these areas and that they are interrelated. Under ideal conditions a treated filler will wet-out and disperse readily in the polymer under conditions of Newtonian flow. The treatment protects the filler against abrasion and cleavage during mixing and in the final composite. The treatment promotes optimum alignment of polymer segments at the interface and overcomes inhibitory catalytic effects the filler may have on polymer cure. The treated filler should remain chemically inert with the polymer during mixing but combine with the polymer during the final cure or moulding operations.[5]

3. RHEOLOGY

3.1. Need for Control

Control of the rheology of filled-resin systems is of such great practical significance in the fabrication of composites that it may outweigh the adhesion promotion of silane coupling agents in importance. Complete dispersion of particulate fillers in the polymer is necessary to eliminate voids or clumps of particles that will act as weak points in the composite. A minimum viscosity is desired in order to incorporate as much low-cost filler as possible, and still obtain adequate flow or handling properties.

Although there have been references to improved processing conditions and increased flow rates attributed to silane modification of mineral-filled rubber,[6] it is only recently that a systematic study has been made on the effect of silane modification of mineral fillers on the viscosity of filled polymers.[7]

3.2. Effect of Water

All hydrophilic mineral fillers and reinforcements have water absorbed on their surfaces which hinders dispersion of the filler in an organic matrix. Water acts as an adhesive to cement ultimate particles together. If the particles are separated by mechanical shear, they reagglomerate in the absence of something to hold them apart.[8]

Dispersion of such fillers in polymers may be improved by pre-drying the filler, or by removing water from the filled compound by milling on heated rolls or by heating in a closed mixer under vacuum. It is very difficult, however, to remove surface water from commercial hydrophilic mineral fillers, and totally impractical to store dry fillers in sealed containers to prevent redeposition of water.

3.3. Dispersion-promoting Additives

It would be very desirable to have additives available that would be effective at low concentrations to give good dispersions of commercial fillers directly into rubber bases. Such a dispersant additive will be an ambi-functional molecule that is compatible with the polymer and has functional groups that can bridge to adsorbed water or even to the filler's polar surface. A convenient way to study dispersion-promoting additives is to measure their effect on the viscosity of filled liquid resins.

Filler/matrix interactions in coating systems were shown to be influenced by acidity or basicity of the filler and resin and of the solvent used to deposit filled-polymer coatings. An acid/base interaction between polymer and filler was necessary for adsorption of polymer. Acidic or basic solvents tend to compete for the polymer or the filler and may assist or prevent adsorption of polymer.[9]

More recently it was shown that minor amounts of polar additives with a compatible, reactive organic group might assist or compete with acid/base interactions in filled, solvent-free polymer systems.[7]

Fillers may be classified as acidic (silica), slightly basic (alumina trihydrate), or basic (calcium carbonate), according to the isoelectric point of the mineral in water. Silane additives may also be described as acidic neutral or basic (cationic) in their reactivity with surfaces:

1. Pre-hydrolysed silane = acidic additive, e.g. $CH_3Si(OH)_3$.
2. Monomeric silane = neutral additive, e.g. $CH_3Si(OCH_3)_3$.
3. Aminofunctional silanes = basic additive, e.g. $H_2NCH_2CH_2NHCH_2CH_2CH_2Si(OCH_3)_3$ (Z-6020®).
4. Titanates and aluminum alkoxides = Lewis acids and catalysts for alkoxysilanes, e.g. $Ti(OBu)_4$, Tyzor® TBT.

3.4. Effect of Additives on Viscosity

Sufficient filler was mixed with mineral oil (prototype of hydrocarbon rubbers) or with silicone fluid (prototype of silicone rubbers) to give a pourable viscosity. The designated additives were stirred thoroughly into portions of the above mixtures and viscosity measured again. The additive effect was calculated as the percentage change in viscosity— either increase or decrease.

Since it is expected that weak van der Waals' forces between hydrocarbons or silicones and fillers are not strong enough to displace water from a hydrophilic surface it is not surprising that addition of polar additives generally caused a marked reduction in viscosity of the filled fluids. The effect of additives on viscosities of three filler dispersions in mineral oil are shown in Table 2.

TABLE 2

VISCOSITY CHANGES IN FILLED MINERAL OIL (%)

0.4% Additive based on filler	27% SiO_2	50% $Al_2O_3 \cdot 3H_2O$	50% $CaCO_3$
None (control), cP	25 000	22 100	40 500
Undecenoic acid	−35	−93	−93
1-decanol	−38	−48	−49
1-aminohexane	−68	−56	−54
$Ti(OBu)_4$	−69	−95	−95
Monomeric $CH_3Si(OCH_3)_3$	−56	−90	−59
Pre-hydrolysed $CH_3Si(OH)_3$	−85	−95	−97
9/1 $CH_3Si(OCH_3)_3$/TBT	−64	−92	−85
Z-6020® (aminosilane)	−76	−70	−79

® Z-6020 is a registered trade name of Dow Corning Corp. (USA).
 Tyzor is a registered trade name of du Pont de Nemours and Co. (USA).

Similar dispersions of fillers in a silicone fluid were modified by polar additives as shown in Table 3. The silicone fluid was an end-blocked polydimethylsiloxane with a viscosity of 500 cP.

TABLE 3

VISCOSITY CHANGES IN FILLED SILICONE FLUID (%)

0·4% Additive based on filler	40% SiO_2	60% $Al_2O_3 \cdot 3H_2O$	46% $CaCo_3$
None (control), cP	90 000	20 000	64 000
Monomeric $CH_3Si(OCH_3)_3$	−75	−16	−93
Pre-hydrolysed $CH_3Si(OH)_3$	−84	+45	−92
9/1 $CH_3Si(OCH_3)_3/Ti(OBu)_4$	−86	−38	−98
Z-6020 (aminosilane)	−95	0	−87
Undecenoic acid	−51	−64	−93
1-butanol	−40	+55	−67
1-aminohexane	−92	−2	−74
$Ti(OBu)_4$	−90	−72	−97

SiO_2 = Minusil® 5, Pennsylvania Glass Sand (USA).
$Al_2O_3 \cdot 3H_2O$ = GHA-332®, Great Lakes Minerals (USA).
$CaCO_3$ = Gama Sperse® LV-10, Georgia Marble Co. (USA).

As expected, almost any polar additive was effective in reducing the viscosity of dispersions of all fillers in mineral oil or silicone fluid. In general, basic additives were best with an acidic filler (silica) while acidic additives were best with basic fillers. Alumina trihydrate in silicone fluid produced some anomalies. Alcohols and some silanes actually increased the viscosity. The best silane modifier was a 9/1 mixture of monomeric methyltrimethoxysilane and butyl titanate. Butyl titanate itself is an effective modifier for control of rheological properties, but in the silane mixture its major function may be to catalyse the condensation of alkoxysilane with water or hydroxyl groups on the filler surface. Alumina trihydrate is also unique in silicone dispersions in that, even without a modifier reasonable viscosities are obtained with relatively high filler loadings.

3.5. Pretreatment of Filler

Pretreatment of filler with alkoxysilanes or mixtures of alkoxysilanes and a catalyst is approximately equivalent to the use of the same modifiers as additives. Butyl titanate is completely ineffective as a pretreatment for silica in silicone fluid even though it is very effective as an additive. When used as an additive, it is probable that only one or two butoxy groups hydrolyse, leaving an oleophilic butoxytitanyl surface modification on the filler. When the filler is pretreated with butyl titanate the treated surface continues to hydrolyse in air with loss of butanol giving a hydrophilic TiO-modified surface. When alkoxysilanes or silane-catalyst mixtures are used in pretreating fillers, the alkoxy groups hydrolyse completely leaving a stable oleophilic organofunctional silicone layer on the filler. Pretreatment of silica with non-catalysed methyltrimethoxysilane is not effective because the volatile monomeric silane is lost before it condenses with the filler, leaving an untreated surface (Table 4).

TABLE 4

VISCOSITIES OF SILICA-FILLED SILICONE FLUID (SILICA PRE-TREATED WITH MODIFIER AND DRIED IN AIR)

Pretreated filler	50% Silica (cP)
None	>100 000
0·5% $CH_3Si(OCH_3)_3$	>100 000
0·3% $Ti(OBu)_4$	>100 000
0·5% 9/1 $CH_3Si(OCH_3)_3/Et_2NH$	7 000
0·5% $(CH_3)_3SiNHSi(CH_3)_3$	5 500

4. FRACTURE

Improved adhesion between resin and filler may not result in improved mechanical performance of composites if stresses across the interface result in filler fracture before the interface fails. Glass fibres are especially sensitive to failure from flaws initiated by scratching against each other during fabrication into composites. One of the major functions of silane finishes is to protect glass fibres against scratching during fabrication.

In composites of rigid resins, it has been observed that silane treatment of fillers with a Moh hardness of 3 or less is ineffective in improving mechanical strength, since cleavage will occur in the filler before failure at the interface.[5]

Silane treatments on clay (Moh hardness = 2·5) have provided significant improvement in mechanical properties of filled rubbers, but it is likely that talc (Moh hardness = 1) would not show similar improvements with silane treatment because of the greater ease of particle cleavage.

5. CATALYSIS

5.1. Cure Inhibition

Fillers are known to have varying degrees of catalytic effect on thermosetting resins but generally inhibit their cure.[10] Inhibition is especially severe in resins cured by free radicals at low temperatures. Inorganic hydroxyl surfaces terminate free-radical reactions by electron transfer to the mineral surface. The resulting mineral free radical is not active enough to initiate another radical chain reaction, resulting in a net termination of chain growth. At temperatures approaching 150°C the mineral free radicals appear to have sufficient activity to initiate new radical chain reactions such that total polymerisation is possible.

Silane treatments on mineral surfaces generally block the inhibitory reactions in both free-radical and epoxide cures. The performance of a silane-treated mineral in a resin composite generally parallels the degree to which the uninhibited reaction is allowed to proceed.

TABLE 5

CLAY-FILLED RUBBER FORMULATIONS

Formulation	Natural (SMR-5)	SBR (1502)	Nitrile (FRN-502)
Rubber	100	100	100
Clay	50	60	70
Zinc oxide	5	4	5
Stearic acid	1·5	2	2
Sulphur	2·5	1·4	0·2
Plastogen®	5·0	—	—
Sunolite®-240	1·5	—	1·0
Thermoflex®-A	1·0	—	—
TMTD	0·5	—	3·0
Santocure®-NS	1·0	—	—
Circolite® oil	—	10·0	—
PBNA	—	1·0	1·0
Flexamine®-G	—	1·0	—
MBT	—	0·8	—
Di-o-tolylguanidine	—	0·3	—
DOP	—	—	10·0

5.2. Effects on Rubber Cure

In some systems (e.g. urethanes and sulphur-vulcanised rubber) certain minerals and certain silane-treated minerals may even catalyse the cross-linking reaction to rates higher than those of unfilled systems.

It is expected that peroxide-cured rubbers will respond to fillers and silane-treated surfaces much like peroxide-initiated polyester cures.

TABLE 6
VULCANISATION CHARACTERISTICS OF RUBBER RECIPES

0·5 % Silane on clay	Mooney viscosity data			Rheometer data	
	viscosity[a]	M3[b]	M18[c]	t_2[d]	t_{90}[e]
Natural					
Control (no Silane)	13	22·6	23·8	4	7·5
Chloropropyl	12	26·7	28·5	7·5	9·0
Mercaptopropyl	17	24·5	26·2	7·5	9·0
Isothiuronium chloride[f]	13	24·9	26·8	7	8·5
Diamine[g]	18	22·4	23·5	7	8·5
SBR					
Control (no Silane)	38	120	149	14	46
Chloropropyl	29	125	161	16	52
Mercaptopropyl	31	103	130	14	48
Isothiuronium chloride[f]	33	101	121	13	33
Diamine[g]	36	94	110	13	34
Nitrile					
Control (no Silane)	36	15·4	22·4	3	11
Chloropropyl	38	16·9	25·1	3	10
Mercaptopropyl	39	15·1	22·5	3	10
Isothiuronium chloride[f]	34	19·8	28·8	3	12
Diamine[g]	45	17·7	25·3	2	12

[a] Viscosity after 4 min at 100°C.
[b] M3 = Mooney scorch at 121°C (min).
[c] M18 = Mooney cure at 121°C (min).
[d] t_2 = Rheometer scorch at 150°C (min).
[e] t_{90} = Rheometer cure at 150°C (min).
[f] Isothiuronium chloride (QZ-8-5456®, Dow Corning Corp., USA):

$$(CH_3O)_3-Si-CH_2CH_2CH_2-\overset{+}{S}-C\begin{smallmatrix}\diagup NH \\ \diagdown NH_2\end{smallmatrix}\ Cl^-$$

[g] Diamine (Z-6020®): $(CH_3O)_3-Si-CH_2CH_2CH_2NHCH_2CH_2NH_2$

Since rubbers are vulcanised at relatively high temperatures the effect of surfaces on rubber cure will not be severe.

Numerous investigators have reported the effects of silane-treated fillers on the rate of cure of sulphur-vulcanised rubbers. Curative adjustments may be necessary for rubbers in which clays, silicas or carbonates are substituted for carbon blacks.[1] Similar modifications may be necessary with a given mineral filler having different silane modifications. Three clay-filled rubber formulations described in Table 5 were examined for scorch times and cure times with four different silane treatments on the clay[11] (Table 6).

An inert silane like 3-chloropropyl generally gave increased scorch time and cure time compared to untreated clay. The 3-mercaptopropyl silane caused less change in cure than the other reactive silanes. The amino-functional silane caused rapid scorch and cure with SBR (containing DOTG accelerator), but was much like the control in natural and nitrile rubbers.

Silica-filled SBR formulations with various curative systems were compared with two silanes added during the mixing operation instead of pretreatment on the filler (Table 7). In all cases, addition of silanes increased the rate of vulcanisation.

TABLE 7

SBR/SILANE RECIPES

Ingredients	Silanes					Controls				
	1	2	3	4	5	1	2	3	4	5
SBR 1502	100	100	100	100	100	100	100	100	100	100
Hi-Sil® 233	60	60	60	60	60	60	60	60	60	60
Z-6062®a	0.9	—	—	0.9	—	—	—	—	—	—
QZ-8-5456®b	—	1.8	1.8	—	1.8	—	—	—	—	—
ZnO	4	4	4	4	4	4	4	4	4	4
Sulphur	1.5	1.5	1.5	1.5	1.5	1.5	1.5	1.5	1.5	1.5
MBTS	1.5	1.5	1.5	1.5	1.5	1.5	1.5	1.5	1.5	1.5
DOTG	1.5	1.5	—	—	—	1.5	1.5	—	—	—
MTD	0.25	0.25	0.25	0.25	0.25	0.25	0.25	0.25	0.25	0.25
Trimene®	—	—	—	1.02	1.02	—	—	—	1.02	1.02
Stearic acid	2	2	2	2	2	2	2	2	2	2
t_2 (scorch), min	4.0	3.0	9.0	2.0	3.5	7.5	5.0	15.0	4.0	4.0
t_{90} (cure), min	10.0	5.0	60.0	8.5	7.0	12.0	9.0	92.5	14.0	10.0

Z-6062® = mercaptopropyl silane.
QZ-8-5456® = Isothiuronium chloride silane.
Vulcanised at 150°C.

6. PERFORMANCE

6.1. In Rubber Compounds

Commercial silane-treated clays were compared by Pinter and McGill[1] with untreated clays in EPDM formulated to 65 durometer. Fillers used in this study are described in Table 8, and rubber properties in Table 9.

TABLE 8

Clay	Trade name®	1978 Price, $/lb
Soft	Paragon	0·0416
Hard	Suprex	0·0364
Delaminated	Polyfil DL	0·1430
Calcined	Polyfil 70	0·1950
Waterwashed, hard	Polyfil HG90	0·1495
Mercaptosilane, hard	Nucap 100	0·1560
Mercaptosilane, hard	Nucap 200	0·1950
Mercaptosilane, water-washed, hard	Nucap 290	0·3380
Aminosilane, hard	Nulok 321	0·2600
Aminosilane, water-washed, hard	Nulok 390	0·3640

® J. M. Huber Corp. (USA)

The silane-modified clays provide higher 200% modulus, lower hysteresis, improved flex and tear resistance with reasonably good tensile strengths, and improved retention of tensile strength after ageing in an oxygen bomb.

The performance of silane modified mineral fillers in rubbers was reviewed by Ranney et al.[12] Fillers may be pretreated with silane, or the silane may be added during mixing of the rubber formulation.

6.2. Importance of Matching

The importance of selecting a coupling agent with an organofunctionality complementary to the curative/polymer is demonstrated in Table 10. In these formulations, 1 phr of each silane indicated was introduced during filler addition on a two-roll mill and the properties were determined by standard methods. In peroxide-cured EPDM formulations, saturated silanes are ineffective due to lack of reactivity in free-radical systems. Unsaturated silanes are effective in the order of their reactivity with free radicals; e.g. a methacryloxypropyl silane is more effective than a vinyl silane.

TABLE 9

CLAYS IN EPDM (65 DUROMETER)

	Paragon® 140	Suprex® 110	Polyfil® DL 110	Polyfil® 70 110	Polyfil® HG90 110	Nucap® 100 110	Nucap® 200 110	Nucap® 290 110	Nulok® 321 110	Nulok® 390 110
Processing										
Mooney scorch, MS3 at 135°C (min)	4	5	5	12	4	4	4	4	4	4
Mooney viscosity, MS4 at 135°C	33	32	22	19	29	31	34	28	29	29
Die swell (%)	85	85	71	90	88	76	77	92	83	84
Rheometer torque at t_{90}, 135°C	39	37	37	36	38	36	38	37	36	36
t_{90} at 135°C (min)	13	13	13	11	12	10	10	11	10	10
Stress/strain (original)										
200% Modulus (psi)										
Cure time 20 min	970	1080	900	620	810	1180	1560	1490	1360	1300
Cure time 30 min	1000	970	860	590	860	1190	1570	1450	1220	1320
Tensile (psi)										
Cure time 20 min	1460	1680	1350	1110	1510	1560	1880	2110	1720	1660
Cure time 30 min	1400	1680	1460	970	1750	1660	1570	1960	1640	1600
Elongation (psi)										
Cure time 20 min	330	320	380	430	380	320	300	280	250	300
Cure time 30 min	300	350	440	410	410	300	300	290	270	300
Flex and tear										
Ross flex (cure time 40 min) (in/100 000 cycles calculated)	0·34	0·17	0·29	0·53	0·15	0·13	0·14	0·23	0·18	0·15
Die C tear (cure time 30 min)										
RT (ppi)	110	153	170	80	168	173	160	133	153	145
100°C (ppi)	65	74	66	54	84	86	96	94	88	92

TABLE 10

RESPONSE OF SILANES AS A FUNCTION OF CURATIVE IN EPDM COMPOUNDS

	300% Modulus, psi	
Silane monomer	Sulphur-cured EPDM, talc filled	Peroxide-cured EPDM, clay filled
Control, no silane	490	420
Amyl	430	410
Vinyl	430	1110
Mercapto	790	1200
Amino	790	1440
Methacryloxypropyl	—	1660

The intermediate performance of mercaptosilanes and aminosilanes is attributed to their activity in chain transfer reactions.

In the sulphur-cured EPDM system, the mercaptosilane and aminosilane appear to be able to participate in the cure mechanism much better than saturated or unsaturated silanes. The mechanism of their reaction in the sulphur cure is not as clear as it is in free-radical cures.

An aminofunctional silane was tested at 1 phr with six different fillers in a polymer blend of EPDM, SBR and NR (Table 11). The silane alone, in the absence of filler, had a modest effect on rubber modulus indicating that it influenced the vulcanisation to provide a higher state of cure. Several fillers showed a similar increase in 300% modulus with added silane, but improvement in tear strength indicated variations in the degree of response. The silica showed greatest improvement, the two clays responded differently, and TiO_2 responded rather well to the silane.

7. ADHESION AND REINFORCEMENT

7.1. Anomalies of Silane Treatment

Although silane-modified minerals show rather creditable properties as reinforcing fillers in various rubbers, the nature and extent of adhesion between rubber and filler remains somewhat of a mystery and clouded by contradictions:

TABLE 11
EFFECTS OF AMINOFUNCTIONAL SILANE

Fillers	300% Modulus, psi			Die C tear, ppi		
	Control	Silane	Improvement, %	Control	Silane	Improvement, %
Polymer blend, no filler	380	490	29	136	146	7
$CaCO_3$	540	810	50	170	183	8
TiO_2	610	950	56	180	224	24
Talc	760	1350	78	187	219	17
Clay, calcined	770	1370	78	180	207	15
Clay, hydrous	940	1660	77	235	301	28
Precipitated silica	730	1300	78	185	240	30

Compound: EPDM/SBR/NR blend, with indicated filler, using guanidine/sulphenamide-type accelerated sulphur cure system. Aminofunctional silane added at 1 phr during milling.

1. Silanes that convert silicates like clay into reinforcing fillers in rubber do not promote good adhesion between similar rubbers and plane mineral surfaces. Silanes that are effective in particulate *resin* composites do improve adhesion of the same resins to plane metal or glass surfaces.
2. Silane treatments on clay that are effective in rubber do not improve mechanical properties of clay-filled expoxies or polyesters. The poor showing of treated clay in rigid resins has been attributed to cleavage within the clay particle when stressed in a rigid matrix.
3. Theoretical concepts of adhesion suggest that the total performance of organic/inorganic composites may be related to polymer morphology that has a minimum in the rubbery range even with optimum chemical modification of the interface.[13]

The apparent contradictions of rubber adhesion as related to rubber reinforcement are resolved if it is recognised that a large concentration of high-energy surfaces may modify adjacent rubber molecules so as to reduce polymer mobility and provide a resinous layer for bonding at the interface. Under such conditions, concepts that have been used in modifying the interface in mineral-reinforced thermosetting resin composites may be applied to mineral-reinforced vulcanised rubbers.

7.2. Recommendations for Other Usage

Limited studies of silanes with vulcanised rubber show enough similarity with thermosetting resins that certain recommendations may be made:[14]

1. A silane should be selected that will combine chemically with the rubber during vulcanisation.
2. Performance will improve with increased crosslinking at the interface.
3. The silane may be applied separately to the mineral or added as an integral blend during compounding of the rubber.
4. Hydrolysable groups on silicon will provide silanol groups for bonding to the mineral, and will be effective on the same minerals that respond to silanes in reinforced plastics.
5. Although only modest improvement in adhesion to vulcanised rubber is provided by thin layers of silanes on plane surfaces, there should be some correlation between peel strengths of vulcanised rubbers on silane-treated glass surfaces and the performance of the same silanes on reinforcing mineral fillers.

7.3. Peel-strength Testing

Several typical rubber formulations were vulcanised against silane-treated glass microscope slides and tested for peel strength. The rubbers were commercial compounds of undisclosed formulation: natural (tyre tread), SBR (tyre tread), nitrile (hose and belts), neoprene (hose, belts and gaskets), EPM and EPDM (sulphur) (ignition-wire insulation) and Hypalon® (ignition-wire jacket).

Microscope slides were dipped in 1% aqueous silanes and dried at room temperature. Non-reactive silanes were selected to contribute varying degrees of surface energy for bonding through dispersion forces.

TABLE 12

ADHESION OF VULCANISED ELASTOMERS TO GLASS (GLASS TREATED WITH 1% AQUEOUS SILANE)

Silane functionality (cure)	Peel strength, ppi		
	SBR (Sulphur)	EPDM	
		(Peroxide)	(Sulphur)
No silane (control)	nil	nil	nil
Non-reactive silanes			
Propyl	nil	0.1	0.7
3-Chloropropyl	0.1	0.1	1.1
Phenyl	0.7	0.2	0.9
Chlorophenyl	0.4	0.2	1.3
Dibromophenyl	0.1	nil	1.5
Carboxyphenyl	0.4	0.2	2.4
Reactive silanes			
Vinyl	0.9	1.1	2.2
Methacryloxypropyl	0.1	2.9	4.6
Aminopropyl	1.1	2.1	5.7
Diamine	1.2	1.3	3.5
Mercaptopropyl	3.0	1.8	8.5
Isothiuronium chloride	6.3	1.0	2.9
Cationic styryl	2.0	20(c)	20(c)

(c) = Cohesive failure in the rubber.

® Hypalon is a trade name of Du Pont (U.K.) Limited.

Sulphur-cured SBR and peroxide-cured EPDM showed slight increases in peel strength with increasing surface energy of the silane-treated surfaces, but none developed significant adhesion (Table 12). Sulphur-cured EPDM showed slightly better adhesion to almost all surfaces.

Reactive silanes were also tested on glass with natural, nitrile, neoprene and Hypalon® compounds.

Chloropropyl-, vinyl-, and methacrylate-functional silanes on glass were only slightly better than the untreated surface for adhesion to most of the rubbers. The mercaptan- and aminofunctional silanes have been the preferred silanes on fillers in sulphur-vulcanised rubbers. The amine gave better adhesion to natural, nitrile and Hypalon® rubbers, while the mercaptan was better with SBR, neoprene and EPDM. The isothiuronium-functional silane appears to be fairly effective in bonding all rubbers—but especially SBR. A cationic styryl coupling agent was the best unsaturated silane and contributed true adhesion to EPDM and Hypalon® (Table 13).

TABLE 13

ADHESION OF VULCANISED ELASTOMERS TO GLASS (GLASS TREATED WITH 1% AQUEOUS SILANE)

Silane functionality	Peel strength, ppi			
	Natural	Nitrile	Neoprene	Hypalon®
No silane	0·1	nil	nil	1·2
Chloropropyl	0·6	0·5	0·1	1·2
Vinyl	0·7	0·1	0·1	1·8
Methacryloxypropyl	0·5	0·1	0·1	2·0
Aminopropyl	1·4	0·1	0·1	6·0
Diamine	4·4	8·0	1·1	20(c)
Mercaptopropyl	2·0	1·2	4·0	2·6
Isothiuronium chloride	3·9	2·0	2·0	2·7
Cationic styryl	4·8	5·5	2·2	20(c)

(c) = Cohesive failure in the rubber.

7.4. Correlation of Peel Strength with Other Properties

A hard kaolin clay (Suprex®) was tested in three rubber formulations as shown in Table 5. Silane-treated clay was dispersed in the rubber on the mill before adding the remaining ingredients. The compounds, with cure characteristics indicated in Table 6, were evaluated for selected mechan-

TABLE 14

MECHANICAL PROPERTIES OF CLAY-FILLED RUBBER (CLAY TREATED WITH 1% AQUEOUS SILANE)

Silane on clay	None	Chloro-propyl	Mercapto-propyl	Isothiur-onium chloride	Amine
Natural					
Peel adhesion, ppi	nil	0·6	2·0	3·9	4·4
300% Modulus, psi	1040	985	1480	1575	1655
Tensile strength, psi	3435	3690	3885	3750	3925
Elongation, %	585	585	540	515	520
Tension set, %	43	43	41	37	38
Compression set, %	8	7	6	5	6
Bashore rebound, %	59	53	64	65	67
Tear strength, ppi	127	138	140	106	118
SBR					
Peel adhesion, ppi	nil	0·1	1·2	1·4	6·3
300% Modulus, psi	285	280	400	405	370
Tensile strength, psi	1120	1380	1505	1590	1720
Elongation, %	925	1015	885	930	975
Tension set, %	37	41	35	36	38
Compression set, %	13	11	10	11	9
Bashore rebound, %	46	47	48	48	48
Tear strength, ppi	141	150	154	162	157
Flex, JIS 10^{-2}	337	567	259	370	520
Abrasion resistance	100	130	155	169	140
Heat build-up, °C	91[a]	103[a]	90[a]	78[a]	97
Nitrile					
Peel adhesion, ppi	nil	0·5	1·2	2·0	8·0
300% Modulus, psi	1230	1355	1755	1770	2125
Tensile strength, psi	3150	3255	3525	4060	3495
Elongation, %	650	665	665	645	545
Tension set, %	31	27	25	28	21
Compression set, %	7	7	6	4	4
Bashore rebound, %	24	23	24	24	24
Tear strength ppi	201	176	167	163	153
Flex, JIS 10^{-2}	49	102	104	75	36
Abrasion resistance	100	122	141	137	166
Heat build-up, °C	61[a]	98[a]	69	57[a]	55

[a] Sample blew out before completion of heat build-up test

ical properties by standard ASTM methods and correlated with microscope-slide adhesion tests (Table 14).[11]

It appears that reaction of organofunctional silanes with rubber types is specific enough that peel adhesion from glass correlates fairly well with mechanical properties obtained with the same silanes on clay filler.

Increased adhesion, as indicated by peel tests, consistently improves certain mechanical properties in all rubbers, has little effect on some properties, and varies with the type of rubber in other properties. Each trend in properties is accentuated by increasing the level of silane treatment from 0.5% to 1.0%.

Increased adhesion is accompanied by

1. 300% modulus: increase;
2. tensile strength: increase;
3. abrasion resistance: increase;
4. tension set: decrease;
5. compression set: decrease;
6. heat build-up: decrease;
7. Shore A hardness: little effect;
8. Bashore rebound: little effect;
9. Mooney viscosity: +natural and nitrile, −SBR;
10. scorch and cure time: +natural and nitrile, −SBR;
11. elongation: +SBR, −natural and nitrile;
12. tear strength: +natural and SBR, −nitrile;
13. flex resistance: +SBR and nitrile, −natural;
14. heat ageing: better natural, poorer SBR and nitrile.

8. ADHESION TO BRASS

Brass is unique among surfaces in providing good adhesion to sulphur-vulcanised rubber without primers or silane treatment. The brass plate must be of optimum composition (about 70% copper), clean, and the rubber-vulcanising system must be adjusted to the brass with respect to reaction kinetics.

The mechanism of adhesion of vulcanising rubber to brass appears to involve reaction of sulphur with both the copper and the rubber. Adhesion is developed even before the rubber cures and involves a resinous crosslinked layer at the metal surface.

8.1. Disadvantages

Even though brass plate is the workhorse in achieving adhesion between rubber and metal, it suffers from some disadvantages. The green tack of rubber compounds to brass is low for rapid fabrication methods, the brass plate is very sensitive to surface oxidation that reduces rubber adhesion, and processing is quite costly.

For this reason brass-plated steel is stored in closed, desiccated containers whenever possible. In addition, thin coatings of synthetic tackifying resins are frequently applied to the surface to protect it against tarnishing. The resin diffuses into the rubber during cure and thus presents a clean surface for the adhesive bond.

8.2. Silane Primers

Silane-modified tackifiers are effective primers for adhesion of thermoplastic elastomers to mineral surfaces,[13] but are generally ineffective with vulcanised rubber. Such primers are not needed to promote initial adhesion of rubber to clean brass, but they may serve some practical functions:

1. They increase green tack with rubber compounds.
2. They protect the surface against tarnishing.
3. They improve the water resistance of the vulcanised rubber/metal bond.

The tack adhesion of unvulcanised skim-stock rubber against primed brass panels is shown in Table 15.

TABLE 15

TACK OF UNCURED RUBBER TO PRIMED BRASS (2% DIAMINE SILANE IN RESIN)

Silane-modified resin in primer	Peel strength after hand pressure, (lb/in)
Unprimed (clean brass)	0·5
Picco® LTP-135	2·3
Piccovar® L-60	3·5
Piccoumarin® 422-R	4·0

® Picco, Piccovar and Piccoumarin are registered trade names of Hercules, Inc. (Wilmington, Delaware, USA).

TABLE 16

ADHESION OF VULCANISED RUBBER TO BRASS (LB/IN PEEL) (TESTED AFTER 6 H WATER BOIL)

Primer on brass	Age of surface before vulcanising (% Cohesive failure)		
	1 h	7 days	10 weeks
None	15·1 (10)	5·6 (0)	1·0 (0)
Z-6020® silane alone	(poor)		
Picco® LTP-135 alone	8·2 (0)		
2% Z-6020® in Picco® LTP-135	29·4 (100)	22 (100)	12 (90)
2% Z-6020® in Piccovar® L-60	23·3 (100)	17 (100)	15 (100)

Adhesion of a commercial rubber skim stock (undisclosed formulation) against brass sheets was examined on fresh surfaces and after ageing in a laboratory atmosphere for up to 10 weeks. Initial adhesion of rubber to brass was generally good, but retention of adhesion after boiling for 6 h in water varied with the history of the brass as shown in Table 16.

Silane-modified tackifier primers on brass not only protected the surface against corrosion that is injurious to adhesion, but improved the water resistance of the bond between rubber and brass, even on clean surfaces.

9. DISCUSSION

Table 14 shows a general correlation of performance of silane-treated reinforcements in rubbers with bulk adhesion, as shown in microscope-slide adhesion tests. However, this is not the complete picture since none of the systems showed very strong adhesion in microscope-slide tests. It may be that reinforcement will not be improved by increasing adhesion over a certain minimum. Mineral surfaces treated with aminofunctional silanes develop strong adhesion to Hypalon® and chlorinated polyethylene[15] rubbers. The cationic styryl-functional silane provides strong adhesion to EPDM rubbers. These combinations should be compared with less-reactive silanes in vulcanised compositions to determine whether really strong adhesion to filler gives superior reinforcement.

Experience with adhesion of rubber to silane-modified primers on brass also suggests that another function of silanes may be to provide long-term resistance to deterioration of rubber properties in hostile environments.

REFERENCES

1. PINTER, P. E. and MCGILL, C. R. *Rubber World*, **2**, 1978, p. 30.
2. WAGNER, M. P. and SELLERS, J. W. *Ind. Eng. Chem.*, **51** (8), 1958, p. 961.
3. BUECHE, F. and HALPERN, J. *J. Appl. Phys.*, **35**, 1964, p. 3142.
4. OBERTH, A. E. *Rubber Chem. Technol.*, **40** (5), 1967 p. 1337.
5. PLUEDDEMANN, E. P. In *Additives for Plastics*, Vol. I, R. B. Seymour (Ed.), Academic Press, New York, 1978, p. 123.
6. GAJEWSKI, M. *Polimery (Warsaw)*, **23** (3), 1978, 97; C.A. 1979, **90**, p. 39972b.
7. PLUEDDEMANN, E. P. In *Additives for Plastics*, Vol. II, R. B. Seymour (Ed.), Academic Press, New York, 1978, p. 49.
8. BURRELL, H. *ACS Div. Org. Coatings and Plast. Chem. Preprints*, **35** (2), 1975 p. 18.
9. MARINO, M. J. Mustofa, M. A., *et al.*, *Ind. Eng. Chem. Prod. Res. Devel.*, **15** (3), 1976, p. 206.
10. PLUEDDEMANN, E. P. and STARK, G. L., *Modern Plastics*, **51**, (3), 1964, 74.
11. PLUEDDEMANN, E. P. and COLLINS, W. T. In *Adhesion Science and Technology*, Vol. 9A, L. H. Lee (Ed.), Plenum Press, New York, 1975, p. 329.
12. RANNEY, N. W., BERGER, S. E. and MARSDEN, J. G. In *Interfaces in Polymer Matrix Composites* (Vol. VI in treatise on composites), L. J. Broutman and R. H. Krock (Eds.), Academic Press, New York, 1974, Chapter V.
13. PLUEDDEMANN, E. P. *Proc. Ann. Conf. Reinf. Plast. Compos. SPI*, **24-A**, 1974.
14. PLUEDDEMANN, E. P. Chapter VI of ref. 12.
15. PLUEDDEMANN, E. P. Unpublished data.

Chapter 6

PLASTICISERS

G. MORRIS
Curtagil Limited, Wokingham, UK

SUMMARY

Petroleum oils are the most widely used extenders and process aids for rubber and related polymers and it is now recognised that the correct choice of material can help the compounder in modifying properties of a component.

With the increasing pressures on all natural resources it is helpful to be aware of the types of material that are available and how their composition influences the processing and characteristics of a polymer and its vulcanisate properties.

A discussion of composition of oils is included, together with a means of determination that can be carried out in an ordinary works laboratory, before indicating the general trends that can be observed when a polymer is compounded with the three main types of oil.

The choice of an oil for a particular vulcanisate property such as low-temperature flexibility or good colour stability is also discussed prior to indicating the specific needs of a few individual polymers that are currently in use.

1. INTRODUCTION

The distinction between materials regarded as softeners, processing aids and extenders under the general heading of plasticisers is not a sharp one. Thus if we consider an oil, a small amount of it may act as a softener and a larger amount as an extender. Today, it is generally accepted that a plasticiser at a loading of 5–20 parts per hundred parts of rubber (phr) is considered to be a process aid and that above 20 phr it is considered to be an extender.

By far the most widely used materials for plasticising natural and synthetic rubbers are petroleum oils which really came into prominence in about 1951 when the first oil-extended SBR masterbatches were introduced in the United States. Mineral oil as an extender for SBR was described in some detail in 1944[1] in one of the many leaflets distributed at the time to those in Great Britain who were interested in the then new SBR. Because the material was not made in this country until many years later, the oil could not be added at the latex stage but had to be incorporated by soaking or milling.

Since 1951, when the cold polymerised rubbers of Polymer Corporation of Canada, and General Tyre Company of USA were introduced, the advantages of low-priced oil-extended grades of SBR have been much appreciated by users, and production has shown rapid increases. The normal grades are mixtures of 37·5 parts of oil with 100 parts of polymer, with the oil being added at the latex stage. Further oil may subsequently be added by the compounder with corresponding increases in black levels to obtain optimum properties.

Oil extension is now also applicable to ethylene–propylene rubbers and also to natural rubber to increase its competitiveness against the synthetics.

2. TYPES OF PLASTICISER

Before the rubber industry realised the benefits associated with mineral oils as plasticisers, many other materials had been considered and some still find a use today. Aside from the petroleum products they can be categorised as pine-tar products, coal-tar products, organic acids, and esters incorporating other synthetic materials.

The pine-tar products are excellent tack producers and have a long-standing use in the manufacture of many articles, including tyres. They give good pigment dispersion and contain phenolic substances which act as antioxidants.

Coal tars were among the first softeners used in rubber but had the disadvantage of being acidic. Certain resins formed from them, for example the coumarone–indene type, had a steady usage and they are credited with improving tensile strength and imparting resistance to cut growth. Their high brittle point and poor light stability are obvious disadvantages.

In the early days of the industry, organic acids were of considerable interest as softeners and plasticisers but now they are only used to activate certain accelerators. For example, a small amount of stearic acid has long been a standard addition to natural rubber mixes, and it serves a similar purpose in most sulphur-vulcanisable synthetics.

The ester-type plasticisers are widely used in polar polymers such as polyvinyl chloride, butadiene–acrylonitrile and neoprene, with PVC creating the largest single demand. They can also be used in less-polar rubbers where they offer good low-temperature properties but they are too expensive for general applications.

3. PETROLEUM OILS

In the early days, petroleum oils were regarded as inferior plasticisers, largely due to the variability of materials offered at that time. However, as the quality improved and the number of products increased it became obvious that they not only aided processing but were also useful in modifying properties.

Rubber compounding oils serve four main purposes;

1. They aid processing of the polymer during milling, mixing and extruding by providing lubrication of the rubber molecules.
2. They improve the physical properties of natural and synthetic rubbers such as elasticity, flex life and low-temperature performance; and also aid the dispersion of pigments resulting in improvements in tensile strength and abrasion resistance.
3. They extend the rubber polymer giving, in effect, a larger volume of elastomer thus off-setting the need for larger production plants.
4. They reduce the cost per pound of the finished rubber goods, providing the cheapest source of raw material in a general rubber compound.

3.1. Types of Oil

Commercial rubber oils are generally high-boiling petroleum fractions obtained after the more volatile petrol and heating-oil fractions have been removed by distillation. They resemble rubbers in that they have a wide distribution of molecular weight but do not follow the normal distribution curve as closely.

There are three principal types of rubber oil, namely aromatic, naphthenic and paraffinic, and each is composed of mainly ring structures. A typical molecule contains unsaturated rings (aromatics), saturated rings (naphthenics) and side chains (paraffinics) (see Fig. 1).

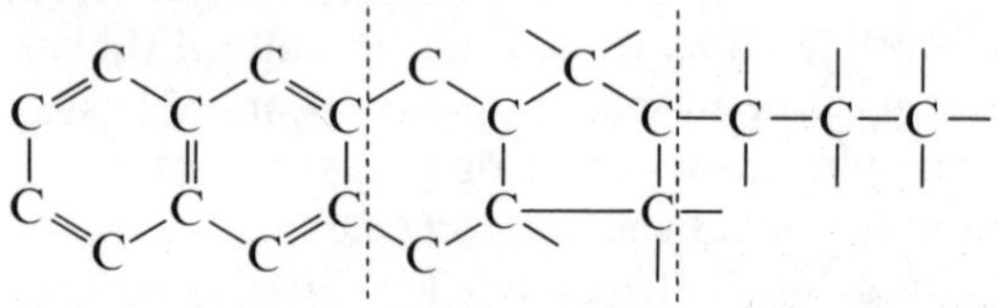

Unsaturated   Saturated   Paraffinic
aromatic rings naphthenic rings side chains

FIG. 1. Typical molecule in oil.

In an aromatic oil there is a preponderance of aromatic ring structures, whilst in a naphthenic oil the predominant ring structures are those containing no double bonds. In a paraffinic oil, the main rings present are again saturated but there are fewer rings per molecule and there are larger numbers of side chains attached. The term 'paraffinic' in this context is really a misnomer since the only paraffins present in a refined oil are waxes.

3.1.1. Aromatics and Paraffinics

These materials occur together naturally, and after removal of light fractions by distillation they are separated by solvent extraction, the aromatic molecules being removed. After removal of the solvent, the aromatic portion can be distilled to give a high-grade aromatic oil or left as a residue containing a high proportion of asphaltenes. The distilled grade will contain very few, if any, asphaltenes but quite a large number of heterocyclics, simply because of structural similarities.

Because they are removed by the solvent treatment, the non-aromatic lube fraction will contain very few heterocyclic structures and no asphaltenes. However, this fraction, which comprises the paraffinic-type process oil, has to be further refined to remove wax which is a very expensive procedure. The total cost of this dewaxing step is not reflected in the price of paraffinic oils because of the value of the wax obtained.

3.1.2. Naphthenics

These oils are produced by distillation of selected naphthenic crudes, ensuring no asphaltenes will be present. Some heterocyclics will be present among the ring structures but naphthenics oils by nature are wax-free.

The types of crudes necessary to produce these naphthenic oils are becoming scarce and this scarcity will, no doubt, exhibit itself in the form of much higher prices in the future. This is already becoming evident, and some of the larger suppliers of rubber oils have only paraffinics and aromatics available.

3.2. Analysis of Rubber Oils

In order to determine the effects of the different types of oil it is obviously important to know what proportion of aromatics, naphthenics and paraffinics are present in a particular grade. To obtain an exact composition is impossible—only a few molecules have actually been isolated. However, from the physical properties of some synthesised high-molecular-weight materials it is possible to postulate what molecules may be present in an oil.

The two methods which are generally used for oil analysis are

1. molecular-type analysis, and
2. carbon-type analysis.

3.2.1. Molecular-type Analysis

This separates the oil into different types of molecules by adsorption on to silica gel or clays, etc., or by chemical means with acids such as sulphuric acid. The procedures are standardised under ASTM D 2006 and 2007 test methods.

These techniques separate the oil into four main groups:

1. Non-hydrocarbon molecules. These contain nitrogen, sulphur, or oxygen and are also called heterocyclics or polar compounds. They have been shown to be responsible for the degradation of oil-extended polymers in storage and under elevated temperature conditions such as in drying operations. It is also suggested that they have an effect on vulcanisation rates.
2. Aromatic molecules. These have more influence on rubber properties than any other molecule and are generally present in the two- or three-ring form. They indicate the compatability of the oil with the rubber.

3. Saturated molecules. The saturates are highly inert, non-polar, and are not removed by the absorbants or acids. Comprising the saturated rings with attached paraffinic side chains they have very good oxidation stability and give good resistance to discoloration by heat and light.
4. Waxes. These should not be present in high-quality oils, but lower-grade materials may have enough present to cause problems of blooming and sweat-out, because of their insolubility in rubbers. Use is made of this property to prevent ozone cracking.

This type of analysis has been used to establish a classification system for extender and processing oils under ASTM D 2226 (Table 1).

TABLE 1

ASTM EXTENDER-OIL CLASSIFICATION

		Asphaltenes, % max	Polar compounds, % max	Saturates, %
Type I.	Highly aromatic	0·75	25	20 max
Type II.	Aromatic	0·5	12	20·1–35 max
Type III.	Naphthenic	0·3	6	35·1–65 max
Type IV.	Paraffinic	0·1	1	65 min

3.2.2. Carbon-type Analysis

The disadvantage of molecular-type analysis is that it does not define the degree of aromaticity or amount of naphthenic character very accurately. For instance, an unsaturated ring structure would appear to be 100% aromatic by molecular-type analysis whether or not any of the carbon atoms were substituted with paraffinic side chains. If side chains were present then compatability with a rubber would obviously be different from that with no side chains present. Carbon-type analysis gives a means of distinguishing these materials by utilising the correlations obtained between the physical properties of pure compounds and hydrocarbon oils containing many types of molecules.

Such a correlation is that based on viscosity/gravity constant, or VGC, and refractivity intercept. This is independent of molecular weight and is based on physical data which is easily obtained in the laboratory

VGC, values of which are now normally given with other typical data of rubber oils, is a measure of aromaticity calculated from the equation

$$\mathrm{VGC} = \frac{10G - 1 \cdot 0752 \log (V - 38)}{10 - \log (V - 38)}$$

where G = specific gravity at 15°C, and
V = Saybolt viscosity at 38°C.
Refractivity intercept is given by the equation

$$\text{Refractivity intercept} = N_D^{20} - 0 \cdot 5 d^{20}$$

where N_D^{20} = refractive index at 20°C for the sodium D line, and
d^{20} = density at 20°C.

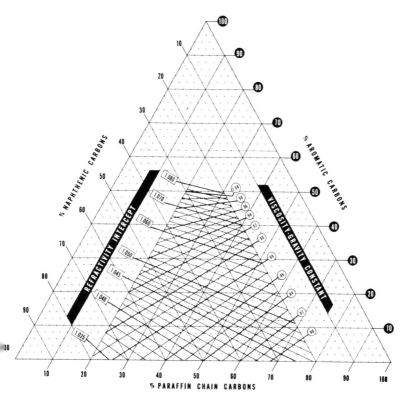

FIG. 2. Triangular graph.

When both these values, which are constant for any given oil, are determined, they are used on a triangular graph (Fig. 2) to indicate the proportion of carbon atoms in the aromatic, naphthenic and paraffinic structures.

This can be applied with reference to Fig. 1. There are a total of 20 carbon atoms, comprising 10 in the aromatic rings (or 50%), 7 in the naphthenic rings (35%) and 3 in the paraffinic side chain (15%). Obviously, this molecule forms a very simple one relating to the aromatic portion of the oil. There will also be present saturated molecules without any aromatic rings, and so the overall composition in terms of C_A, C_N and C_P will be expressed as a percentage of the total number of carbons present in both the saturated and unsaturated fractions.

Thus carbon-type analysis gives an estimate of the degree of aromaticity and it can be obtained from a few physical tests which are simple to perform. This method and molecular-type analysis are normally quoted in typical test data of rubber oils and both are useful to the compounder in comparing oils from different sources.

Typical properties of the three types of rubber oil that are currently commercially available are given in Table 2.

TABLE 2

TYPICAL PROPERTIES OF RUBBER OILS

	Paraffinic	*Naphthenic*	*Aromatic*
Viscosity (cSt)			
At 40°C	19·7	110·2	763·5
At 100°C	4·0	8·0	17·0
Specific gravity at 15°C	0·861	0·932	1·018
VGC	0·809	0·885	0·980
Refractive index	1·4751	1·5167	1·5804
Refractivity intercept	1·0457	1·0503	1·0721
Carbon-type analysis			
C_A	3·5	21	45
C_N	31·0	37	18
C_P	65·5	42	37
Molecular-type analysis, weight %			
Asphaltenes	0	0	0
Polar compounds	0·4	2·8	7·8
Aromatics	12·1	42·8	80·0
Saturates	87·5	54·4	12·2
Aniline point, °C	96·0	75·0	38·2

4. COMPOUNDING EFFECTS OF RUBBER OILS

The main use of petroleum oils in rubber is to aid processing, but the type of oil used can also modify the physical properties of the finished product. Therefore consideration has to be given to the processability of the raw or compounded polymer, the strength properties of the compound which give some measure of its quality, and the elastic properties of the product which give an indication of what may be expected under dynamic conditions.

The following general trends are those recorded on SBR polymers since they form the main bulk of synthetic materials used, but they may be considered to hold good in the main for the other polymers in general use, some of which are highlighted later.

4.1. Processability

Processing is defined as any compounding step commencing with mixing and ending at vulcanisation, and thus covers Banbury mixing, sheeting, calendering and extrusion.

The addition of oil at the mixing stage serves to reduce the overall viscosity of the mix to a workable level, reduce power consumption at high filler loadings, keep heat generation down and ease the dispersion of pigments.

These are important criteria since they are contributory to reducing mixing time, increasing production and reducing power consumption.

The time taken for an oil to form a cohesive mass with SBR is reduced as the aromatic content of the oil increases and its molecular weight decreases (Fig. 3).

When fillers are also included in the mixing cycle the incorporation time for proper dispersion becomes important. For example, the dispersion of carbon black is influenced by the amount of oil used, the sequence in which the oil and black are added, and the composition of the oil. Increasing the amount of oil decreases the degree of dispersion. This effect, which is more pronounced if the oil is added too early in the mixing cycle, has been attributed to the decreased viscosity of the stock. To minimise this effect, the oil and filler should be added alternately to avoid drastic softening.

At any given oil level, the degree of dispersion of the black is influenced by the type of oil used. It has been reported[2] that aromatic oils give the best dispersion, followed by naphthenic and paraffinic materials in that order. It was also evident that HAF black gave the best dispersions and SAF black the worst (Fig. 4).

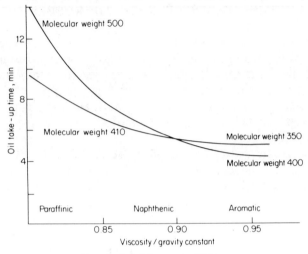

Fig. 3. Oil take up in SBR.

The same trend was observed whether the oil was added on an equal weight or an equal volume basis, although the effect was less marked where equal volumes were used.

This effect, of the greater solvent power of the aromatic oil giving better dispersion, can possibly be attributed to the structural similarities

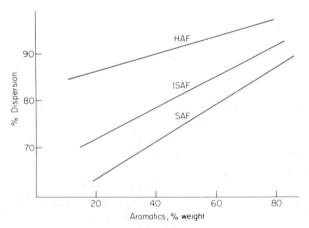

Fig. 4. Dispersion versus oil aromatics.

of the black and oil molecules. It is proposed that this similarity aids the solvation of the carbon-black particles, prevents them from forming agglomerates, and enables them to be carried in the oil throughout the rubber, giving better dispersion.

Good dispersion with the more saturated oils can be obtained with longer mixing but this increases the cost of the product. However, it may be necessary since the dispersion of pigments and fillers determines the optimum physical properties of a compound, as will be seen later.

After mixing, oils play an important part in extrusion by increasing extrusion rates, giving smoother stocks and regulating die swell. They are also effective in such operations as calendering, where close tolerances must be maintained and in moulding where good flow characteristics are necessary.

With regard to cure rates, which are always important to the rubber compounder, it is found that aromatic oils give the fastest cures at least during the early stages. This may be due to the non-hydrocarbon, sulphur and nitrogen content of these oils, or just the greater polarity of the aromatic molecules over the saturated ones, although this has not been established.

2. Vulcanisate Properties

Although oils are added to rubber primarily to aid processability, they also affect the physical properties of the compound after vulcanisation and here again oil composition together with its loading and viscosity are important variables.

2.1. Oil Loading

Oil loading, for any given filler loading, determines the hardness of a compound and in turn, depending on the proportions added, will affect other physical properties. A compromise is necessary to obtain a good overall balance. A point to note when comparing oils of different composition is that they will have different specific gravities and at any given weight loading, their volumes may vary quite substantially. This is all the more important when dealing with oil-extended polymers and some of the newer rubbers which are capable of accepting very high loadings. At any given weight loading, the lower-gravity material will be present in greater volume, and to eliminate loading effects in comparing oils it is desirable to use equal volumes of each.

4.2.2. Oil Viscosity

Oil viscosity, which increases with increasing molecular weight, also has an effect on rubber properties particularly in regard to low-temperature performance and minimising losses of oil at high temperatures. When considering low-temperature performance, the important parameter is the viscosity of the plasticiser at the temperature in question. This in turn is again influenced by oil composition. Aromatic oils show the greatest change of viscosity with temperature, and are said to have the lowest viscosity index, whereas paraffinic oils show the least change of viscosity with temperature and have the highest viscosity index. Thus, assuming there are no problems of incompatibility, paraffinic oils will impart best low-temperature performance and the lower-molecular weight or lower-viscosity materials will be most effective. Conversely high-temperature losses of oils can be minimised by using higher molecular-weight or higher-viscosity oils. The flash point of an oil serves as a useful guide to its volatility characteristics and to minimise losses during production of rubber compounds, oils with flash points below about 200°C should be avoided. To minimise oil losses from cured compounds, and hence maintain desired physical properties of a compound, the compatibility of oil and polymer is extremely important. The greater the affinity of the oil for the rubber, the less likely it is to migrate to the surface of the compound and be removed. This situation becomes more critical under dynamic test conditions. To satisfy these requirements, again assuming no problems of compatibility, the paraffinic oils, especially the higher-molecular-weight materials, give the best results. This is very evident in the highly loaded EPDM compound where, for maximum retention of physical properties when a compound is subjected to high temperatures or dynamic test conditions, paraffinic oils are used almost exclusively. For the unsaturated polymers, e.g. polychloroprenes, polybutadienes and nitriles, the high-viscosity aromatic oils give the best results, principally due to their compatibility with these materials.

In regard to other viscosity effects, it is generally observed that for a series of oils of the same type added to a rubber, the strength properties of the vulcanisates increase with an increase in oil viscosity while resilience decreases.

It is also generally noticed that, with a series of oils of the same type, compatibility with a polymer tends to decrease as molecular weight or viscosity of the oil increases.

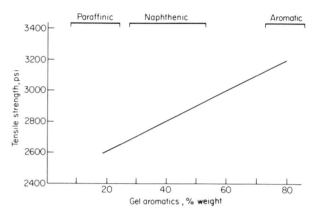

Fig. 5. Tensile strength versus oil aromatics.

4.2.3. Oil Composition

In addition to the effect that the viscosity of an oil has on vulcanisate tensile strength, there is also the effect that its composition can impart. Tensile values increase as the aromaticity of the oil increases (Fig. 5) and this is attributed to the better dispersion of carbon black that is obtained with this material. With sufficient time to get the required dispersion, high tensile strengths can also be obtained with the more saturated rubber oils.

Modulus and elongation at break are only slightly affected by oil composition. With increasing aromaticity, modulus tends to decrease and elongation tends to increase.

Hardness is affected even less by oil composition which at first seems rather surprising as one would expect greater softening by the more aromatic materials. This may be due to the volume effect mentioned previously; namely, the higher volumes of the less aromatic materials that are necessary to give a constant weight loading counterbalance the effect of the greater softening of the more aromatic oil.

Conversely, tear strength is greatly influenced by oil composition in oil-extended compounds. Significant increases in tear resistance are found with increasing aromaticity (Fig. 6) and it is suggested that this property may again be related to better carbon-black dispersion.

Oil composition can show an effect on the dynamic properties of a rubber compound such as resilience, heat build-up and crack growth.

Paraffinic oils confer the lowest heat build-up (Fig. 7) due to their

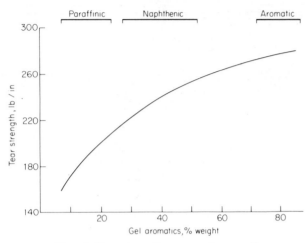

FIG. 6. Tear strength versus oil aromatics.

better lubricating properties and also give the best rebound resilience, as would be expected. They are not as good as aromatics, however, in imparting resistance to crack growth, where an almost 100% improvement can be noted in the De Mattia test values in going from 16% to 83% aromatics in oil composition (Fig. 8). In this test a nicked sample is used and so this is more a measure of tear resistance which has been discussed above.

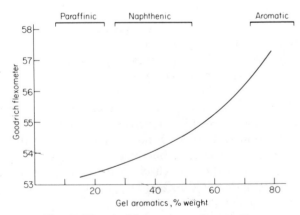

FIG. 7. Heat build-up versus oil aromatics.

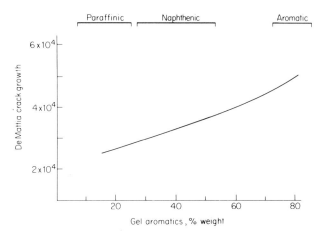

FIG. 8. Crack growth versus oil aromatics.

5. OIL STABILITY

With the broadening of the applications of synthetic polymers and a corresponding increase in their performance requirements have come greater demands on such properties as heat stability, ageing characteristics and staining tendencies. These demands obviously increase the performance requirements of the plasticisers used in them, and petroleum technology has had to keep abreast of these changes.

The heat and light stability, ageing characteristics and staining tendency of an oil are all linked to its oxidation stability and the best oxidation stability is shown by the saturated paraffinic materials. It was originally thought that instability was directly related to an oil's aromatic content, particularly its highly condensed ringed systems. It is now known that oxidation instability of an oil is due primarily to the presence of nitrogen and sulphur heterocyclics in the oil, and that the saturated naphthenic portion and polynuclear aromatics with as many as four and five condensed rings are stable. However, since these heterocyclics tend to be concentrated in the aromatic portion of the oil (see Section 3.2), it is easy to see why the aromatic oils show the worst oxidation stability. Tests which measure aromatics do give a measure of oil instability since they measure that component which contains the offending heterocyclic molecules. It is not practical or economical to produce aromatic oils free from heterocyclics and so aromatic oils which improve processing generally, are not recommended where heat or light stability, good ageing or non-staining properties are required.

Generally, naphthenic oils shown oxidation properties intermediate between the paraffinic and aromatic oils but a premium will have to be paid to obtain these materials in the years ahead as they become more scarce.

6. USE IN SPECIFIC POLYMERS

6.1. Ethylene–Propylene Polymers

The introduction of EPM and EPDM polymers placed more emphasis on oil stability than any other series of materials. With their outstanding resistance to ozone and weathering, high heat resistance and good electrical properties, they demanded similar performance from any plasticiser used in conjunction with them. This is all the more so with EPDM which is capable of accepting high loadings of oil and filler.

The copolymers are peroxide-cured and are used solely with highly saturated oils, i.e. the paraffinics, because of this. The non-hydrocarbon components of an oil which are present in larger quantities in the aromatic oils account for acceleration depletion and the corresponding adjustment in peroxide level has to be made.

The terpolymers are much more widely used, and the introduction of the diene to enable them to be sulphur-curable permits them to be extended with a wide range of process oils. Loadings of up to 100 parts of oil and corresponding increases in black loading give only small decreases in Mooney viscosity, and even higher loadings give Mooney values within workable range. Some EPDM polymers are available which accept oil loadings over 200 phr.

To maintain their weathering characteristics and high heat resistance these polymers are again generally compounded with a high-molecular weight, high-viscosity paraffinic oil.

6.2. Natural Rubber

Natural rubber is used in the manufacture of many products and compounds may be made satisfactorily with most types of petroleum oil. The increasing competitive pressure of synthetic rubbers, with which it is often used in combinations, forced a reassessment of the possibility of oil-extending natural rubber in the 1960s by Moore et al.[3] This work showed that with the proper choice of mixing conditions and the addi ion of extra black to compensate for the added oil, no serious impairment of physical properties occurred in the unaged or aged compounds up to about one-third by weight of oil. In a tyre compound, tyre wear for the oil-extended grade was shown to be about 15% better than with the unextended control under severe test conditions and about 10% better under moderate conditions. Overall tyre wear was shown to be similar to oil-extended SBR tyres run in the test programme.

These results seem to show that this technique can be used to make natural rubber fully competitive with SBR and oil-extended SBR. It remains to be seen whether it can be viable economically, however, since in some countries it loses its status as a raw material when it is extended in the latex stage, and suffers the penalty of import taxes.

5.3. Butyl Rubber

Butyl rubbers are copolymers of isobutylene and isoprene in which unsaturation is kept at a low level, which accounts for butyl's excellent ageing and resistance to ozone cracking combined with low permeability to air and other gases. Because of their high internal viscosity, butyl rubber vulcanisates have low resilience at room temperature, and this internal viscosity rises with decreasing temperature, resulting in a leathery state being reached at $-18°C$. To increase the elasticity, additional crosslinks need to be introduced or the internal viscosity can be reduced by the use of low-viscosity oils. This moves the minimum of the curve obtained when rebound is plotted against temperature to a lower temperature (Fig. 9). For comparison the rebound curves of natural rubber and unplasticised butyl rubber are shown.

Lack of volatility of the plasticiser at higher temperatures is equally important and to satisfy both these requirements, paraffinic oils are preferred, since, as we have seen, they exhibit the highest viscosity indexes as well as showing the better compatability with butyl.

6.4. Polychloroprenes

The original use of oil in polychloroprenes was to give practical processing properties in the uncured stock and there was little need to be too selective although oils with a VGC of about 0·885 were preferred.

The WHV grade of neoprene was developed to take high loadings of oil and filler and is used for packings and mechanical goods where tensile strength is not too important. The oils used should not bloom and those with high aromaticity are essential as indicated by reference to the amount of oil absorbed as a function of aniline point (Fig. 10).

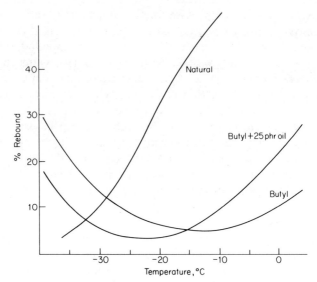

Fig. 9. Effect of oil on rebound.

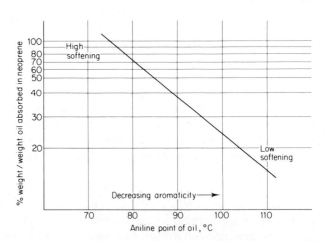

Fig. 10. Oils as softeners for neoprene.

6.5. Nitriles

The use of process and extender oils in nitrile rubbers is quite limited but when they are used checks should again be made on blooming and in general the more aromatic oils should be used for compatability.

More commonly, however, ester plasticisers are used though there is scope to use the more highly refined oils either solely or in conjunction with the esters to show economic advantages.

6.6. Liquid Elastomers

The introduction of liquid elastomers that are easily converted to solid rubbers has revolutionised traditional methods of rubber manufacture.

Again oil has been shown to offer a means of reducing compounding cost and modifying liquid-elastomer properties and again in general the aromatic content of the oil determines compatability with the individual chemical components and the cured liquid elastomers. Thus aromatic oils show the maximum compatibility but paraffinic oils may still be used. They require higher temperatures to give the necessary solubility with the components and are only recommended at low loadings with the cured elastomer.

In general, all types of oil can be used and the selection of any particular type may well depend on the final elastomer characteristics required, production economy and other considerations.

REFERENCES

1. *The Oil Technique for GRS*. Manufacturers Memorandum No. M.12, The Services Rubber Investigations, London, April 1944.
2. STOUT, W. J. and EATON, R. L. Compounding oils. *Rubber Age*, December 1967.
3. MOORE, C. G., et al., *Rubber Developments*, **15** (2), 1962.

Chapter 7

PROTECTIVE AGENTS

B. T. ASHWORTH and P. HILL
Vulnax International Limited, Manchester, UK

SUMMARY

The current situation on antidegradants mainly as it applies to vulcanised rubbers is reviewed.

Some indications of the size and division of the market for these products are given, and preferred chemical types are detailed, together with the background to their choice and adaptability, in sections headed staining antioxidants, non-staining antioxidants, staining antiozonants and non-staining antiozonants.

Sections are also devoted to synergism, antioxidant permanence and health and safety aspects.

1. INTRODUCTION

The rubber industry is a well-established and mature industry and, as a result, changes in the pattern of development and usage of compounding ingredients occur relatively slowly. During the last decade, for instance, there has been little discovered in the way of new antidegradant structures: new products which have appeared commercially have been variations and optimisations on well-known themes. Similarly, there have been no major contributions to the general understanding of the frequently listed main causes of oxidative rubber deterioration—heat, light, flexing, ozone and metal catalysis. The last major area of uncertainty was that of ozone degradation which received enormous attention and clarification in the period 1955–1965.

The interest which has existed and the progress which has been made in recent years surrounds topics such as permanence of antioxidant effects and health and safety factors. Physical form has received some attention as part of an increasing need to improve working conditions in rubber factories. This issue, however, is less worrying in the case of antioxidants than it is with accelerators since many antioxidants, particularly the most widely used staining products, lend themselves to presentation in clean, dustless bead or flake forms, on account of their melting points and chemical nature.

A major review of antidegradants for rubber, with more than a hundred references was carried out by Leyland[1] in 1972 for the then Institution of the Rubber Industry. Those seeking detailed information should start with this document. A much shorter but comprehensive and thoroughly readable survey of antioxidants and antiozonants in rubber compounding was published by Parks and Spacht in 1977.[2]

2. INDUSTRY STATISTICS

Antioxidants and/or antiozonants are essential to the performance of most rubber compositions and generally they represent the greatest cost amongst the low-dosage, high-effect chemicals used.

An approximate idea of the size of this market can be obtained from world rubber consumption. This was of the order of 12 million tonnes in 1977 and it is forecast by the World Bank to reach 24 million tonnes in 1990, with the main growth of consumption taking place in the developing countries and the centrally planned economies. Assuming a conservative dosage of 1 part per hundred of rubber of antidegradants, a world-wide consumption of at least 120 000 tonnes is indicated at present by the rubber compounding industry. The consumption of aniline-based (i.e. amine-based) antioxidants in the USA and Western Europe was estimated in 1975.[3] Abstracted data are given in Table 1.

These data indicate, as might be expected, that p-phenylenediamines are the group of antidegradants in greatest demand. The absence of dialkyl-p-phenylenediamines from the table is surprising.

The 1980 predictions for phenyl-β-naphthylamine (PBN) and phenyl-α-naphthylamine (PAN) look unduly pessimistic now. Substantial amounts are still used and there seems no reason to suppose that this will change radically in the short term. A more recent estimate puts Western European demand for staining antidegradants by the rubber

TABLE 1

AMINE ANTIOXIDANTS AND ANTIOZONANTS: PRESENT AND FUTURE CONSUMPTION IN THE USA AND WESTERN EUROPE (TONS)

	USA		Western Europe	
	1974	1980	1974	1980
Diaryl-p-phenylenediamines	6 750	7 000–7 250	2 100	2 900–3 000
Alkyl/aryl-p-phenylenediamines	19 500	21 500–22 500	17 000	21 000–22 000
PBN and PAN	2 200	nominal	5 000	nominal
Dihydroquinolines	4 000	6 500–7 000	5 500	6 500–7 000
Diphenylamine condensates and miscellaneous products	15 000	15 500–16 000	6 750	9 600–10 000
	47 450	50 500–52 750	36 350	40 000–42 000

TABLE 2
STAINING (DISCOLOURING) ANTIOXIDANTS: RELATIVE PERFORMANCE IN NR

Composition	Heat resistance	Flexcracking resistance	Metal inhibition	Ozone resistance	Comments
Phenyl-β-naphthylamine	Good	Good	Good	None	General-purpose antioxidant once widely used for stabilising synthetic rubber and in compounding where discoloration not important. Recently encountered toxicity problems (see Section 9).
Phenyl-α-naphthylamine	Good	Good	None	None	Used mainly as an antioxidant for oils and greases. Recently encountered toxicity problems (see Section 9).
Acetone/diphenylamine condensation product—solid	Very good	Moderate	None	None	Used in cables, general mechanical goods and those parts of tyres where heat-ageing protection is the main requirement.
Acetone/diphenylamine condensation product—liquid	Very good	Very good	None	None	Popular antioxidant for use in tyres, re-treading compounds and heavy mechanical goods such as conveyor and transmission belting. Excellent antioxidant for nitrile rubber (NBR) vulcanisates. Also used for stabilisation of synthetic rubber

PROTECTIVE AGENTS 231

dihydro-2,2,4-trimethylquinoline					conjunction with a *p*-phenylene diamine derivative. Also used in cables and general mechanical goods where heat ageing is the prime requirement. Useful antioxidant for peroxide cures. Powder often used in latex for carpet-backing compounds. Very low staining characteristics.
Octylated diphenylamine	Good	Moderate	None	None	Excellent, indeed best, single antioxidant for polychloroprene. Least staining of amine antioxidants.
Alkylated diphenylamines (heptyl, octyl, nonyl)	Good	Moderate	None	None	Largely used in latex or oil industries. Available in self-emulsifying forms for use in latex. Very low staining characteristics.
Diphenyl-*p*-phenylene-diamine	Not assessed because of very low compatibility				Used as a secondary antioxidant in NR/SBR to improve flexcracking resistance. Used in polychloroprene in blends to improve ozone resistance.
Di-β-naphthyl-*p*-phenylenediamine	Very good	Weak	Excellent	None	Powerful inhibitor of metal-catalysed oxidation. Useful latex antioxidant. Also exhibits a high degree of permanence. Has the same problem as PBN.

processing industry at 30 250 tonnes in 1978.[4] Demand for non-staining products is also estimated for the same period at 4000 tonnes.

A further very important market for antioxidants is in synthetic elastomers stabilisation. These products are added to the polymer during manufacture to protect it during drying and subsequent transport and storage before use. World production of synthetic rubber in 1978 amounted to 8·7 million tonnes[5] requiring of the order of 80 000 tonnes of antioxidant products for stabilisation. Synthetic rubber production in Western Europe in 1978 was close to 1·9 million tonnes, representing a market of around 20 000 tonnes for stabilisers.

3. STAINING ANTIOXIDANTS

Virtually all of the staining antioxidants currently in use were discovered during the 1930s. By far the most important groups of these are based on aniline and diphenylamine. Indeed, all of them can be considered to be derived from aniline since diphenylamine itself is an aniline derivative. The napthylamines—and in particular PBN which until recently enjoyed widespread use—have declined in popularity because of doubts concerning their toxicity (see Section 9). Where PBN has been replaced, the alternatives have been either acetone/diphenylamine condensates, in cases where flexcracking as well as heat protection was considered important, or by acetone/aniline condensates (the dihydroquinoline derivatives) where the emphasis was mainly on heat resistance. The latter, in combination with *p*-phenylenediamine derivatives, is widely used in tyre compounding.

For the stabilisation of synthetic rubber, the use of PBN has been virtually eliminated. Phenolic products such as butylated hydroxytoluene (BHT) and styrenated phenol have proved popular as alternatives. The main staining antioxidants used commercially at present are listed in Table 2.

4. STAINING ANTIOZONANTS

The major advance in knowledge of the factors governing ozone attack of rubbers and ways in which the attack could be eliminated or minimised took place in the period 1955–1965. This progress was stimulated by more-exacting service requirements in articles such as tyres and by the increasing use of synthetic rubbers, particularly styrene/butadiene rubber (SBR).

Hydrocarbon rubbers containing chemical unsaturation in the polymer main chain or backbone may be attacked by ozone when vulcanisates are in a condition of strain. If the strain is essentially a static one, valuable protection may be obtained from waxes or flexible lacquers. Under conditions of dynamic strain, however, these remedies are less effective and can even become prodegradant. Under dynamic service conditions the use of chemical antiozonants is essential for long-term, balanced protection.

4.1. p-Phenylenediamine Derivatives

All major commercial antiozonants are derivatives of p-phenylenediamine and correspond to the structure given below (**I**).

$$R_1-NH-\underset{I}{\underset{}{\bigcirc}}-NH-R_2$$

Two classes are of commercial interest:

1. The symmetrical p-phenylenediamines where $R_1 = R_2 =$ alkyl, cycloalkyl or substituted aryl.
2. Asymmetrical products where $R_1 =$ aryl (most usually phenyl) and $R_2 =$ alkyl or cycloalkyl.

The products of greatest technical interest are listed in Table 3. 6-Ethoxy-2,2,4-trimethyl-1,2-dihydroquinoline is noteworthy because it is the only commercially available structure capable of both static and dynamic ozone protection which is not a p-phenylenediamine.

The symmetrical dialkyl-p-phenylenediamines, as illustrated by B(DMP)PD and DCPD in Table 3, give superior protection under static exposure conditions whereas the asymmetrical structures, exemplified by IPPD, 6PPD and CPPD, are better for dynamic protection. The inferior static protection given by the asymmetrical products can be much improved by the addition of waxes, however, and therefore this class has an overall superiority over the symmetrical types. Furthermore, most of the alkyl-substituted symmetrical products have the drawback of liquid physical form.

TABLE 3

SOME COMMERCIALLY AVAILABLE ANTIOZONANT STRUCTURES

	Melting point, °C	Mooney scorch at 120°C, min[a]	NR Protection		SBR Protection		De Mattia flexing, kc[b]
			Static	Dynamic	Static	Dynamic	
Alkyl/aryl-substituted p-phenylenediamines							
N-Isopropyl-N'-phenyl-p-phenylenediamine (IPPD)[c]	75	23	100	100	100	100	430
N-1,3-Dimethylbutyl-N'-phenyl-p-phenylenediamine (6 PPD)	45	26	100	90	85	80	300
N-Cyclohexyl-N'-phenyl-p-phenylenediamine (CPPD)	118	24	130	80	85	80	300

Structure							
CH₃-CH-CH₂-CH₂-CH₂-CH-N... (see name below)	Liquid	19	160	65	130	75	200
N,N′-Bis(1,4-dimethylpentyl)-p-phenylenediamine [B(DMP)PD]							
(dicyclohexyl structure)	104	15	150	65	125	70	225
N,N′-Dicyclohexyl-p-phenylenediamine (DCPD)							
Diaryl-p-phenylenediamines							
(ditolyl structure)	95	25	50	70	50	40	280
N,N′-Ditolyl-p-phenylenediamine (DTPD)							
Others							
(ethoxy-trimethyl-dihydroquinoline structure)	Liquid	21	45	50	50	60	230
6-Ethoxy-2,2,4-trimethyl-1,2-dihydroquinoline (ETQ)							

[a] Mooney scorch at 120°C — the times quoted are those observed for a rise of 10 Mooney units above the minimum value. The result was obtained in an NR tyre treadstock with 2% of the antiozonant. The value obtained with the blank was 27 min.
[b] De Mattia flexing — the results quoted were obtained in an NR tyre treadstock with 2% of the antiozonant. A 3-in free sample length was used, and the results quoted are the number of kilocycles required to produce deep cracks. The result obtained with the blank was 87 kc.
[c] IPPD was taken as standard and given a rating of 100.

The diaryl-*p*-phenylenediamines are powerful antiozonants for polychloroprene rubbers (CR) and are often favoured because of their minimal effect on bin storage. DTPD has been claimed to provide long-term protection of general-purpose rubbers[6] though as shown in Table 3 initial activity is modest.

Choice of specific structure within each of these two classes depends not only on differences in antiozonant efficiency between individual products but also on factors such as their effect on processing safety, on physiological factors such as skin sensitisation and, of course, on price. Undoubtedly the most powerful and versatile chemical antiozonant is IPPD closely followed by 6PPD. There has been a slow trend over the years for 6PPD to replace IPPD on account of factors such as the reduced tendency of 6PPD both to cause skin sensitisation and to be subject to water leaching, although the practical significance of the latter factor has been questioned.[7]

Technically, structures like IPPD represent the ultimate in antidegradant performance available at the present time since, in addition to their ozone resistance, they also give protection comparable with the best available in heat resistance, flexcracking protection and metal inhibition.

5. NON-STAINING ANTIOXIDANTS

Non-staining antioxidants were mainly discovered in the period 1945–1955. Those of greatest interest to the rubber industry are nearly all derived from phenols: they fall mainly into two chemical classes:

1. Simple hindered phenols—usually trisubstituted in the 2, 4, 6 positions.
2. Hindered phenols bridged in the *ortho* or *para* position by a methylene group, substituted methylene group or sulphur.

5.1. Simple Hindered Phenols (II)

$$\text{structure II: phenol with } R_1 \text{ (ortho), } R_2 \text{ (para), } R_3 \text{ (ortho)}$$

II

R_1, R_2 and R_3 are either alkyl, cycloalkyl or aralkyl. The substituents may be the same, as in styrenated phenol, but are more generally different and are most frequently a mixture of alkyl groups. Apart from styrenated phenol, the most commonly encountered products are methylated in the *ortho* and/or *para* positions and contain either t-butyl, octyl or nonyl groups in the remaining position(s). Very few of these products, the main exception being 2,6 di-t-butyl-*p*-cresol (BHT), are single chemical entities: most of them are complex liquid mixtures.

In technical performance, simple hindered phenols are usually inferior to bridged hindered phenols in protection against heat ageing. However, certain members of the former group show unparalleled protection against deterioration by light and flexing.[8] Simple phenols are generally more volatile and more easily extracted than the bridged ones—the latter feature leading to a tendency to stain adjacent materials.

5.2. Bridged Hindered Phenols

This group of non-staining antioxidants contains those offering the highest general levels of protection. By far the majority of the products are bisphenols possessing structures of the type **III** or **IV** below (R_1, R_2 and R_3 are either H, alkyl or cycloalkyl). An important exception is the dicylopentadiene-bridged structure (**V**).

5.3. Antioxidant Efficiency

Antioxidant efficiency in vulcanised rubber is related to structure in a well-understood manner, the highest activity being obtained with bisphenols which are *ortho* bridged. It must be stressed that this observation applies only to performance in rubber vulcanisates and not in other polymeric materials where in some instances considerably higher levels of activity are shown by the *para*-bridged compounds.

Within the more rubber-effective, *ortho*-bridged structures, the following additional structural considerations apply:

1. nature of the bridge structure:
2. effect of the hindering group;
3. effect of the *para* substituent.

5.3.1. Nature of the Bridge Structure (X)

Maximum protection is obtained in the structures where bridging is by carbon, single carbon bridges being preferred $(X = CH_2)$. Increasing the bridge from one to two carbon atoms has little influence on antioxidant performance but the tendency of the resulting chemical to impart pink colorations to rubbers is reduced. Further increases in the length of carbon bridges have an unacceptable adverse effect on activity.

In the case of one-carbon bridges, substitution of one hydrogen atom by alkyl improves activity, except where the two phenol elements are *ortho* substituted by t-alkyl or cycloalkyl. Other substituents for hydrogen in the one-carbon bridge, such as phenyl, cycloalkyl, chloro or trichlormethyl groups, reduce activity markedly. Substitution of both hydrogens on a methylene bridge also reduces activity markedly.

Alternative bridging atoms such as sulphur $(X = S)$ also result generally in poor activity but such a structure is useful in the case where R_1 is t-butyl.

5.3.2. Effect of Hindering Group (R_1)

Alkyl and cycloalkyl groups are the most effective *ortho* substituents. Chloro or nitro groups destroy activity. Methyl-substituted compounds are widely encountered but peak activity is obtained when t-butyl or α-methylcyclohexyl substituents are used.

5.3.3. Effect of para-Substituent (R_2)

Quite high activity is obtained when the *para* position remains unsubstituted but the tendency of compounds to discolour is increased. Best activity is obtained with a methyl group but this can result in pinking with most t-alkyl or cycloalkyl hindering groups. Use of an ethyl group in the *para* position reduces the tendency of the compound to pink albeit with a noticeable reduction in activity.

Maximum activity in this class of compounds is, therefore, obtained when the bridging group $X = CH_2$, $R_1 =$ t-butyl or α-methylcyclohexyl and $R_2 =$ methyl.

An informative study of the relationship between performance and chemical structure in polynuclear phenols has been made by Kempermann.[9]

5.4. Factors Affecting Choice

So many non-staining antioxidant structures are available commercially that any attempt to list them would result in a lengthy catalogue. The compounder's task to select the most appropriate product is not an easy one. In addition to their sheer numbers, the products vary widely in the different aspects of antioxidant protection offered, and furthermore the differences between any two products can and do change according to factors such as the elastomer or the vulcanisation system used. In addition to the immediate technological performance there are factors such as permanence, degree of acceptance for foodstuffs or pharmaceutical use, physical form, price and extent of staining and discoloration to be considered. Unless an enormous amount of experimental work is to be carried out, the compounder must rely in the first instance on information provided by the supplier, although he should always satisfy himself on very specific aspects of his service requirements such as migration staining or contamination of contact fluids.

6. NON-STAINING ANTIOZONANTS

The search by rubber chemicals manufacturers for a non-staining antiozonant providing powerful protection under both dynamic and static conditions in the general-purpose rubbers, NR and SBR, has still not been successful. It is probable that the research effort in this area has been reduced over the last decade, not only on account of the lack of useful leads but because of the more widespread use of ozone-resistant rubbers, particularly EPDM. Trends in component design—particularly in car windscreen surrounds—have also reduced the need for such a chemical compared with the situation in the 1960s.

6.1. Commercial Products

Non-staining antiozonants have been introduced but their areas of successful application are limited. Tributyl thiourea has proved capable of giving non-staining ozone protection but there can be problems with bloom and effects on cure. Indeed the ozone protection obtained may well be due to the bloom providing a protective layer on the surface of the vulcanisate in much the same manner as wax.

Over the last decade, Bayer[10] have introduced three interesting products in this area. These are novel structures having ether linkages in common. One is a non-staining antiozonant for CR and its blends with natural rubber, under static conditions, which also exhibits antioxidant activity in both NR and CR. The second is also an antiozonant for CR but it is effective in other general-purpose rubbers under static exposure conditions if used in conjunction with wax. The third displays the same qualities as the second but to a greater extent. Neither of the latter two structures confer antioxidant protection.

6.2. Waxes

Waxes are well-known materials for conferring non-staining ozone protection under static conditions of exposure. An elegant account relating the choice of wax to climatic conditions has been given by Morche.[11]

6.3. Rubber Blends

Although not fitting into the generally accepted idea of an antiozonant it should be recalled that the ozone resistance of general-purpose rubbers can be improved by blending with an ozone-resistant rubber such as EPDM. Compared with chemical antiozonants, the amount required is high, usually 10–30%. The improvement to ozone resistance obtained is very dependent on the grade selected and the physical properties of the vulcanisate, and it is highly sensitive to the completeness of blending of the two elastomers and their compatibility in terms of the vulcanisation system used.

7. SYNERGISM

Many examples of antioxidant synergism in polymers exist but the number of genuine examples observed in vulcanised rubbers seem remarkably small. There are many cases in which a desirable spectrum of protection against several types of deterioration is achieved by blending two or more products whose individual protection features supplement each other. There are others where an appropriate combination of a powerful and a weak antioxidant provide a protective system with a attractive cost/efficiency balance. However, true synergism—where the level of a given facet of performance of an antioxidant combination is raised above that achievable by using any of the constituents on their own—seems to be confined to combinations of phenolic or amine antioxidants with 2-mercaptobenzimidazole (MBI) or closely related derivatives. These synergistic blends are capable of giving at least three types of protection:

1. against metal-catalysed oxidation;
2. against heat ageing, particularly with low-sulphur cures;
3. against flexcracking in certain rubbers.

The synergistic systems, comprising MBI and an amine or phenolic antioxidant, usually contain the two components in equal proportions by weight. There is some evidence, however, that the optimum ratio is dependent on the curing system employed: best results are obtained, for example, in peroxide-cured ethylene/propylene copolymer (EPM) with a 4 : 1 ratio of MBI to amine antioxidant.

7.1. Alternatives

Since at least some of the aspects of MBI synergism have been known for over 20 years, it is surprising that no other structures have been commercialised which fulfil a similar function. There are indications that MBI is being replaced by a mixture of its 4- and 5- methyl derivatives at the present time on account of the better availability of the chemical intermediate for the latter product.

In the absence of vulcanisation some additional possibilities for synergism exist. Thus in thermoplastic styrene/butadiene/styrene (SBS) rubbers it has been shown that antioxidant systems of the type well known to the plastics industry, consisting of powerful phenolic antioxidants, together with chemicals such as dilauryl thiodiproprionate (DLTP) or tris(nonylphenyl)phosphite (TNPP) give both excellent protection and clear synergism.[12] Probably the best systems in this area have still to be optimised.

8. PERMANENT ANTIOXIDANT EFFECTS

A great deal of the development work and concern in the antioxidant field in recent years has been concentrated around the permanence of the effects obtained. The reasons for this are at least threefold:

1. A desire to improve product performance by reducing antioxidant loss by volatilisation at high temperatures or by extraction by solvents, detergents or steam distillation.
2. A need to conserve antioxidant in the rubber for economic reasons.
3. A recognition of the nuisance and possible health hazards arising from antioxidant vapours in the atmosphere.

8.1. Improving Permanence

There are basically two ways of obtaining a substantial improvement in antioxidant permanence in rubber:

1. Selecting or creating a high-molecular-weight antioxidant.
2. Reacting the antioxidant with the rubber.

8.1.1. Selecting or Creating a High-molecular-weight Antioxidant

A good example of a long-established antioxidant exhibiting excellent resistance to volatilisation and extraction by a wide variety of liquid media is N,N'-di-β-naphthyl-p-phenylenediamine. The best way, clearly, to achieve maximum effectiveness using this approach is to use polymeric products: certain acetone condensates with aniline or diphenylamine have been in the form of low polymers for many years and they show greater permanence than many other products. The use of polymeric antioxidants made by reacting hydroquinone or p-benzoquinone with diaminoaromatic amines, for instance, has been discussed.[13] Products of this type, however, have not been commercialised.

8.1.2. Reacting the Antioxidant Chemically with the Rubber

In this approach the antioxidant is bound to the polymer backbone—perhaps the best-known method is that developed by workers at the Malaysian Rubber Producers' Research Association (MRPRA)[14] involving the reaction of C-nitroso compounds with unsaturated rubbers as shown in Fig. 1. The best-known example of this class of compound is p-nitrosodiphenylamine (NDPA). When this compound is used an N-phenyl-N'-substituted p-phenylenediamine is formed where the N' substituent is the rubber hydrocarbon. The resulting protection against heat ageing is comparable with that of IPPD and the effect resists exhaustive solvent extraction. Unfortunately no protection against flex-cracking or ozone is obtained possibly due to the lack of antioxidant mobility. NDPA as such presents serious handling problems and attempts were made both by MRPRA[15] and ICI Ltd[16] to overcome this

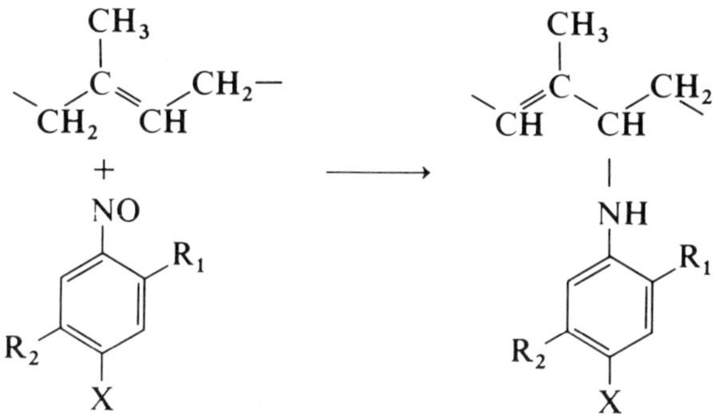

FIG. 1

Most commonly when $X = -NH-\text{C}_6\text{H}_5$, R_1 and $R_2 = H$

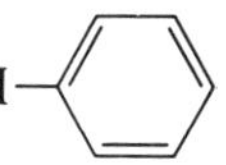

and when $X = -OH$, R_1 and $R_2 = H$ or alkyl.

Fig. 1

problem by forming adducts which regenerated NDPA during vulcanisation. Other problems, however, prevented successful commercialisation.[17]

MRPRA[18] have also shown that N,N'-disubstituted quinone diimines and N-substituted quinone imines can react with rubber to give the corresponding p-phenylenediamine or p-aminophenol, together with a rubber-bound entity.

Grafting effects to latex have also been described.[19] Antioxidants containing vinyl groups such as 3,5-di-t-butyl-4-hydroxybenzylacrylate (**VI**) can be successfully grafted with an appropriate redox-initiating system. Furthermore, certain products without vinyl groups can be grafted in this way,[19] an essential requirement apparently being the presence of a methylene grouping *para* to the phenolic group. Thiol-containing anti-

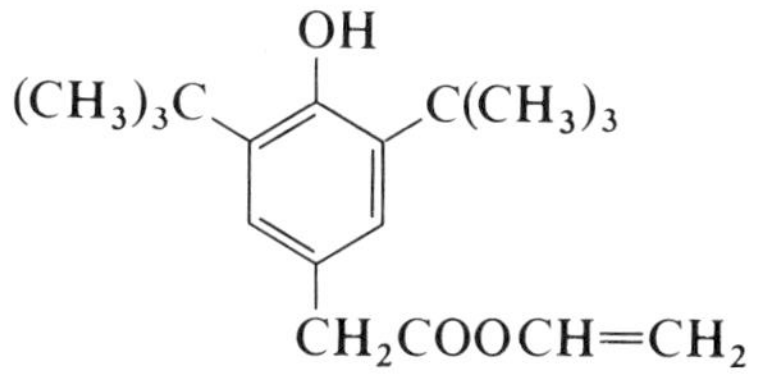

VI

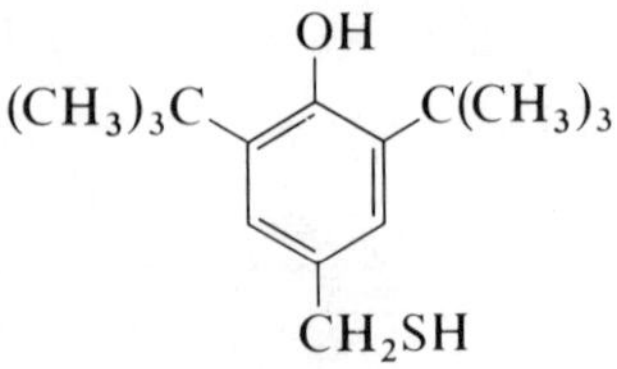

$$\underset{\text{VII}}{\underset{\text{CH}_2\text{SH}}{\text{(CH}_3)_3\text{C}}\diagdown\overset{\overset{\text{OH}}{|}}{\bigcirc}\diagup\text{C(CH}_3)_3}$$

VII

oxidants such as 3,5-di-t-butyl-4-hydroxybenzyl mercaptan (**VII**) are also claimed to give high yields of bound antioxidants in latex; they are apparently readily added to the double bonds in rubber in the presence of free radicals.

Interesting and imaginative as these ideas are, none has yet been exploited commercially.

In the case of synthetic rubber, however, Kline and Miller and co-workers have shown that N(4-anilinophenyl) methacrylamide (**VIII**) may be incorporated into SBR or NBR during polymerisation to give a rubber-bound antioxidant.[20,21]

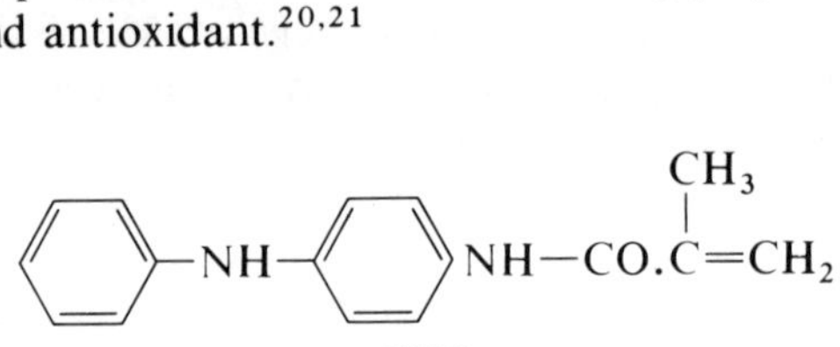

$$\bigcirc\text{—NH—}\bigcirc\text{—NH—CO.}\overset{\overset{\text{CH}_3}{|}}{\text{C}}\text{=CH}_2$$

VIII

NRB rubbers based on this principle have been commercialised and claimed to give superior high-temperature performance to conventional grades.

The preoccupation with extraction effects may be seen in developments with conventional antioxidants also. Two papers[22,23] provide data showing that thermal embrittlement of NBR in mineral oil is not only delayed by the presence of certain antioxidants in the rubber, but also further delayed by the presence of similar antioxidants in the oil, which reduce extraction effects from the rubber. A certain substituted phenol has been shown to be capable of imparting antioxidant effects to NBR which resist solvent extraction[17] and this effect has been used both in compounding and in NBR stabilisation. This performance cannot be explained in simple network-binding terms.

The effect of a selection of antioxidants on the microbiological attack of natural-rubber pipe rings immersed in water has been studied.[24] It was concluded that antioxidants were beneficial in proportion to the extent to which they resisted extraction by water, with the exception of the network-bound *p*-nitrosodiphenylamine which did not reduce microbiological attack at all.

Systematic studies on the vaporisation of antioxidants from vulcanised rubbers would be useful to have. The general class of antioxidant which is worst in this respect is almost certainly the substituted phenols. A recent analysis of the oily condensate on the inside of a tyre-factory window showed it to be almost entirely composed of BHT. This puzzled the chief chemist, as this product was not used in the factory, until it was realised that it originated from the synthetic rubber.

9. HEALTH AND TOXICITY

The most significant issue in recent years, affecting both the thinking and practice, in antioxidant usage or, for that matter, in rubber compounding as a whole, has surrounded the well-known and long-established substituted naphthylamines. The products involved are PBN (IX), PAN (X) and N,N'-di-β-naphthyl-*p*-phenylenediamine (DNPPD) (XI). Most of the debate and study has been concerned with PBN which in volume terms was by far the most important to the rubber industry. Commercial grades of these chemicals, as supplied for many years from a variety of sources, contain small quantities—parts per million—of the known carcinogen β-naphthylamine as impurity. Despite the lack of positive evidence from epidemiological studies connecting the use of PBN with a higher-than-normal incidence of bladder cancer[25,26] and regardless of the introduction of a new grade containing less than 1 ppm of free β-naphthylamine,[27] the use of the product in both compounding and synthetic-rubber stabilisation has declined. It is interesting to reflect, however, that the view has been expressed that because of all the probing into the use of PBN which has taken place in recent years, there are unlikely to be any surprises remaining in connection with its use—a degree of confidence which may not exist for many popular but less-studied products.

PAN and DNPPD have declined in popularity with the compounder along with PBN. Whilst PAN never had any unique technological niche, DNPPD was valuable in that it not only gave impressive protection against heat and metal catalysis initially but that protection had outstanding resistance to volatilisation and extraction by liquid media.

Another major concern surrounding rubber chemicals in recent years, connected primarily with health and safety, is that of physical form. The more serious issues, however, concern accelerators. Whilst there is room for improvement in some areas of antioxidant presentation, their nature, melting points and methods of manufacture generally permit a presentation in bead, rod or flake forms, produced from the molten condition, which possess clean, dust-free characteristics.

Antioxidants used in rubbers intended for foodstuffs contact, and as stabilisers in synthetic rubbers to be used for this purpose, will obviously be required to show appropriate approval under the Food and Drug Administration (FDA) regulations of the USA and/or the Bundesgesundheitsamt (BGA) regulations of West Germany. In the future it is felt that this type of approval will be requested increasingly by end-users in applications where no foodstuffs contact is involved, as additional general reassurance concerning product handling.

Responsible manufacturers of rubber chemicals have been providing information on potential hazards and correct handling of their product for many years, but there are clear indications that the amount of such information is increasing and that it is becoming more detailed. For those who may prefer such advice from an independent source, the British Manufacturers' Association (BRMA) has published a document entitled *Toxicity and Safe Handling of Rubber Chemicals*, BRMA Code of Practice 1978.

REFERENCES

1. LEYLAND, B. N. *Progr. Rubber Technol.*, Vol. 36, PRI, 1972, p. 19.
2. PARKS, C. R. and SPACHT, R. B. *Elastomerics* (5), 1977, p. 109.
3. BATTANI, N. D. *Chemie Developpment International, European Chemical News*, 21 November 1975, p. 36.
4. Information Research Ltd. *Chemical Age*, 20 October 1978.
5. Rubber Statistical Bulletin, Vol. 33, No. 8, May 1979. International Rubber Study Group, London.
6. WIDMER, H. W., SHUTTLEWORTH, M. J. and COLLONGE, J. *Intern. Rubber Conf.*, *Kiev*, October 1978.
7. BROWNING, G. R. and BARNHART, R. R. *Paper No. 25, ACS Div. Rubber Chem.*, Miami Beach, April 1971.
8. *Technical Data 1, Permanax WSL and Permanax HO*, Vulnax International Ltd, Manchester, UK.
9. KEMPERMANN, TH. *Technical Notes for the Rubber Industry*, No. 45, 1972, Farbenfabriken Bayer AG, Leverkusen, West Germany.
10. *Vulkanox AFC, Antiozonant AFD and AFS*, Farbenfabriken Bayer AG, Leverkusen, West Germany.
11. MORCHE, K. *Rubber Plastics Age*, **48** (9), 1967, p. 1094.
12. ASHWORTH, B. T. and HILL, P. *Fourth Australian Rubber Technol. Conv.*, October 1977.
13. KAY, E., THOMAS, D. K. and WRIGHT, W. W. *Intern. Rubber Conf.*, Brighton, May 1972.
14. CAIN, M. E., KNIGHT, G. T., LEWIS, P. M. and SAVILLE, B. *Rubb. Res. Inst. Malaysia*, **22**, 1969, p. 289.
15. British Patent 1 340 672.
16. British Patent 1 379 232.
17. ASHWORTH, B. T., LEYLAND, B. N. and QUAN, P. M. *Intern. Rubber Conf.*, Brighton, May 1972.
18. CAIN, M. E., GELLING, I. R., KNIGHT, G. T. and LEWIS, P. M. *Rubber Ind.*, **9** (6), 1975, p. 216.
19. SCOTT, G. *Intern. Rubber Conf.*, Brighton, May 1977.
20. KLINE, R. H. and MILLER, J. P. *Rubber Chem. Technol.*, **46** (1), 1973, p. 96.
21. MEYER, G. E., KAVCHOK, R. W. and NAPLES, F. J. *Rubber Chem. Technol.*, **46** (1), 1973, p. 106.
22. BLOW, C. M. *Rubber J.*, **155** (7), 1973, p. 18.
23. YUROVSKII, V. S., MALYSHEV, A. I., PETROSYON, N. E. and KIM, Z. K. *Kauch. i Rezina* **6**, 1975, p. 38.
24. CUNDELL, A. M., MULLCOCK, A. P. and HILLS, D. A. *Rubber J.*, **155** (4), 1973, p. 22.
25. British Rubber Manufacturer's Association, Health Research Unit, Birmingham. Bulletin No. 9, October 1971.
26. FOX, A. J., LINDARS, D. C. and OWEN, R. *Brit. J. Ind. Med.*, **31** (2), 1974, p. 140.
27. *Technical Information R196*. ICI Ltd, Organics Division, 1971.

Ideally product design calls for a detailed knowledge of the physical properties of the commonly available rubbers together with expertise in manufacturing processes and a sound knowledge of engineering principles. It is perhaps not too surprising that designers of rubber products are not always as familiar with the properties of rubbers as they should be for optimum performance at minimum cost. The object of this chapter is to survey the general properties of rubbers from the designer's point of view and then to show how other factors, such as shape, can often permit the use of a rubber which does not appear to meet the original requirements in terms of general physical properties. A good example of this is the use of non-oil-resisting rubbers for automotive engine and gearbox mounts.

Another area where lack of experience may lead to incorrect conclusions is the interpretation of standard test results. It must always be borne in mind that standard tests are, as the name implies, tests carried out in a standard way. They are not necessarily the most informative tests that could be carried out for a particular application. It is very important to keep in mind the exact nature of the information being sought, and then to decide whether or not the test under consideration is likely to provide the relevant data. A noteable example is the laboratory testing of abrasion resistance which is not used by tyre companies because experience shows that such tests do not predict tyre wear.

In the case of accelerated tests which are normally carried out at high temperature there is a difficulty. In some cases raising the temperature does accelerate the test but in others quite different physical or chemical effects become dominant at the higher temperature so the results are meaningless. The reason for using higher temperatures to represent longer times is that observations show that both physical phenomena associated with viscoelastic behaviour and the rate of chemical reaction occur more rapidly as the temperature is raised. Where more than one effect is involved both may well increase with temperature but a crossover may occur so that one effect dominates at one temperature while the other is more important at a different temperature. In extreme cases the phenomenon may not occur at all at higher temperatures so it is very important to understand the mechanism involved. For example, as will be seen later, ozone protection is conferred by wax blooming which is due to the presence of more wax than is soluble at room temperature. Raising the temperature, besides speeding up the rate of diffusion to the surface, increases the solubility of the wax in the rubber so that eventually it becomes completely soluble and no more blooming takes place

Clearly in this instance raising the temperature does not indicate what would happen at longer times.

When considering standard test results it is important to note the values of similar or related tests in order to obtain a reliable conclusion. A specification may call for a relatively small change in hardness after accelerated ageing at high temperature because a stable network is required. However, changes in hardness arise from two opposing effects—main-chain scission and/or crosslink breakage, both of which reduce hardness and additional crosslinking from unreacted vulcanisation residues which increase hardness. These two effects may cancel so that the net result is zero but compression-set tests which are also carried out at elevated temperatures would show a large amount of set due to the crosslinks which have been formed in the compressed sample. Under these conditions a truly stable network would not be affected.

The component designer cannot be expected to possess the detailed knowledge of compounding of the rubber technologist but he should be aware of the general principles as outlined below.

2. SELECTION OF RUBBER COMPOUND

2.1. Introduction

In designing a component it is usually necessary to compromise since many desired properties are mutually exclusive. However, some properties may be varied considerably without serious detriment to others and a knowledge of these is essential if a satisfactory selection is to be made. There are, of course, many physical properties to consider and it is only practicable to discuss the more important ones in this chapter.

Probably the single most important property and certainly the most widely measured is hardness. The hardness of a rubber compound can be changed in three ways:

1. by the introduction of a filler;
2. by the introduction of an extender or plasticiser;
3. by varying the number of crosslinks formed during vulcanisation.

The next most widely specified property is tensile strength and this may be surprising at first sight because rubbers are hardly ever used in applications where the deformation is tensile and even less where the strains are near to the breaking strain. What then is the purpose of

2.4. Vulcanisation

The third method of changing the hardness is to vary the amount of crosslinking in the vulcanisate. This is most easily achieved by changing the amount of crosslinking agent in the rubber. There are many crosslinking agents but most use sulphur together with accelerators to speed up the process. The choice of a particular crosslinking system depends on many factors but the degree of crosslinking (i.e. the number of crosslinks per unit volume of rubber) is easily varied. The larger the number of crosslinks the harder the compound will be. However, as mentioned before, the tensile strength normally passes through a maximum so the 'optimum' cure is normally the one giving the maximum tensile strength. Increasing hardness by crosslinking leads to a higher resilience compound unlike the effect of adding fillers.

It can be seen that hardness can be varied in many ways but the best way depends on the importance of the other properties which are also influenced by the method chosen to vary hardness.

2.5. Low-temperature Crystallisation

Rubbers which strain-crystallise on stretching also crystallise at low temperatures but in this case it takes a considerable time before crystallisation effects are noticeable.[4] The effect of the crystallisation is to stiffen the rubber considerably and the only way of removing it is to raise the temperature. The rate at which crystallisation occurs depends on the temperature, rising at first as the temperature is lowered and then after passing through a maximum it falls with decreasing temperature. The temperature for maximum crystallisation rate in natural rubber is $-26°C$ and in polychloroprene $-5°C$. The crystallisation rate is more rapid in uncrosslinked rubbers and is also influenced by the type of crosslinking system used.[4] Low-temperature crystallisation is not a problem if the exposure is relatively short term, say a few days at a time or if the component is subject to frequent movements since the heat generated during these movements will melt the crystals. However, components in a 'static' application and exposed for long periods to low temperatures will crystallise if made from strain-crystallising rubbers. It is just this strain-crystallising behaviour which makes these rubber strong at normal temperatures so if a strong rubber is required which will not crystallise at low temperatures it will be necessary to choose a copolymer or other anticrystallising rubber and to rely on reinforcing fillers to provide adequate strength at normal temperatures.

2.6. Low-temperature Flexibility

The behaviour of rubbers at low temperatures (except for crystallisation as mentioned above) depends on the glass transition temperature of the rubber. All rubbers become stiffer as this temperature is approached and the resilience passes through a minimum value. The glass transition temperatures of common rubbers are given in Table 1 and it can be seen that for low-temperature work the choice is limited to silicone rubber, butadiene rubber and natural rubber.

TABLE 1

GLASS TRANSITION TEMPERATURE, MAXIMUM NORMAL WORKING TEMPERATURE AND RELATIVE COST OF VARIOUS RUBBERS

Rubber	Glass transition temperature, °C	Maximum normal working temperature, °C[a]	Relative cost
Silicone	−120	200	12
cis-Polybutadiene (BR)	−108	70	1·2
Natural (NR)	−70	70	1
Butyl (IIR)	−65	90	1·3
Styrene/butadiene (SBR)	−61	70	1·2
Polychloroprene (CR)	−49	90	2·6
Butadiene/acrylonitrile[b] (NBR)	−24		1·8
Fluorocarbon		200	35

[a] See text for cases where these temperatures are exceeded in service.
[b] 38·5% acrylonitrile.

All rubbers can be improved to some extent in their low-temperature behaviour by the incorporation of a plasticiser. This is a low-viscosity, low-volatility liquid (e.g. dioctyl adipate or dioctyl sebacate) which is added to the rubber during the mixing stage and will, depending on the amount incorporated, extend the acceptable temperature range to some 10 or 20 degrees lower than the non-plasticised rubber.[5] The degree of crosslinking does not affect this behaviour significantly; neither does the presence of a filler although blending with a rubber with a lower glass transition temperature can be beneficial.

2.7. Viscoelasticity

All rubbers show viscoelastic behaviour which means that their physical properties are partly liquid-like (viscous) and partly solid-like (elastic). For example, the thermal expansion coefficients of rubbers are similar to those of liquids not solids. In fact rubbers behave in many ways like very viscous liquids before they are crosslinked. The vulcanisation process which introduces crosslinks reduces the flow properties and makes a rubber more elastic. Nevertheless, there is still evidence of flow behaviour even in a crosslinked rubber. This is demonstrated in the creep and stress-relaxation behaviour of rubbers.

Those unfamiliar with these two properties are probably aware of the time-dependent nature of the hardness measurement which necessitates the specification of the time interval between applying the load and taking the reading in the standard hardness test. The pointer on the machine is often still moving after the 30-s period has elapsed indicating that the indentor is still slowly sinking further into the rubber. This is clearly not elastic behaviour in the classical sense which demands an instantaneous response to a deforming force. Nevertheless, for practical purposes recovery will be complete if sufficient time is allowed to elapse after the load is removed. Note that a period of time several times longer than the time for which the load was applied is necessary to ensure complete recovery.

This time-dependent behaviour arises from the viscous nature of the polymer and is minimised by increasing the degree of crosslinking which improves the elastic part of the behaviour. The type of crosslink is also important. Conventional sulphur vulcanisates are the worst in this respect although these crosslinks give highest strength. Direct carbon–carbon crosslinks produced by organic peroxide crosslinking agents are much better in reducing the time-dependent properties to a minimum but do not produce such high tensile strengths. The same is true of the efficiently vulcanised sulphur compounds (low sulphur, high accelerator) which contain mostly monosulphide crosslinks.

If it is more important to minimise the time-dependent effects such as creep and stress relaxation than to have the highest possible tensile strength then carbon–carbon crosslinked or monosulphide crosslinked rubbers should be used in preference to conventional sulphur crosslinked rubbers.[1] An additional benefit has been found recently by using fully soluble compounding ingredients for the monosulphide crosslinked rubbers so that these compounds are just as good as the carbon–carbon crosslinked material.[6]

2.8. Oils and Solvent Resistance

There are no rubbers that are completely resistant to mineral oils but the more polar the rubber the less oil will be absorbed. Polychloroprene is the most rubbery polymer at room temperature which is reasonably oil-resistant. Greater resistance is shown by acrylonitrile/butadiene co-polymer and the greater the proportion of acrylonitrile the better is the

TABLE 2

RESISTANCE OF RUBBERS TO ORGANIC LIQUIDS†

Liquid	NR	SBR	IIR	CR	NBR	FKM	CSM
Acetone	S	S	S	D/U	U	U	S
Benzene	U	U	F/D	U	U	S	—
Carbon tetrachloride	U	U	U	U	F/D	S	—
Castor oil	F	F	S	S	S	—	S
Chloroform	U	U	U	U	U	—	U
Coconut oil	U	U	S	S	S	—	S
Corn oil	U	U	S	F/D	S	—	S
Cotton-seed oil	U	U	S	S	S	—	S
Creosote oil	U	U	U	F	S	—	—
Diethyl ether	U	U	F	U	S	—	S
Ethyl alcohol	S	S	S	S	S	—	S
Ethylene glycol	S	S	S	S	S	S	S
Hexane	U	U	U	S	S	S	S
Kerosene	U	U	U	F/S	S	S	F/S
Lubricating oils	U	U	U	S	S	S	S
Methyl ethyl ketone	U	U	F	U	U	U	—
Olive oil	U	U	S	F/S	S	—	S
Trichloroethylene	U	U	U	U	U	—	U
Vegetable oils	U	U	S	F	S	—	—
Xylene	U	U	U	U	U	S	—

Key:
- S Satisfactory for general use.
- F/S Satisfactory although resistance is not a maximum.
- F Fair, but probably satisfactory.
- F/D Medium resistance—control tests recommended.
- D Doubtful—control tests strongly recommended.
- D/U Low resistance.
- U Unsatisfactory.

Extracted from *Rubber in Chemical Engineering* by S. Buchan (published by Plant Lining Group, Federation of British Rubber and Allied Manufacturers, London, and MRPRA, Hertford, 1965) where a more comprehensive range of liquids is given.

oil resistance. However, these materials are intermediate between true rubbers and flexible plastics. Their resilience is low, their elongation at break is relatively low, they are relatively stiff for rubbers even without fillers. However, if really good oil resistance is required for seals or for hoses these are the materials used. Other rubbers may be used if the oil is of vegetable origin, as shown in Table 2.

The same principle applies to solvents: non-polar solvents are absorbed by non-polar rubbers but not by polar rubbers whereas polar solvents are absorbed by polar rubbers but not by non-polar rubbers. Intermittent contact with solvents may lead to the loss of mobile materials from the rubber (e.g. antioxidants, extending oils, etc.) by a leaching process even if the solvent does not swell the rubber a great deal. Care should be taken to ascertain which liquids are likely to come into contact with a rubber component in order to decide on the correct material to use. However, as already discussed, the time required for a significant amount of liquid to be absorbed should be estimated since this may show that a rubber which is resistant to the liquid is unnecessary because the component will have the required lifespan even if continuously immersed in the liquid.

2.9. Heat Resistance

The maximum temperature to which a rubber may be exposed without suffering a substantial loss in physical properties depends on a number of factors such as time of exposure, thickness of rubber, and availability of oxygen, so it is difficult to decide on a generally applicable value. Typical figures used by technologists as a guide to the worst situation (i.e. readily available oxygen usually from the air and thin test piece continuously exposed) are shown in Table 1. In contrast it is worth noting that there are applications where these temperatures are exceeded in actual service conditions without detriment mainly due to the lack of oxygen in the region of high temperatures. Truck tyres which are normally natural rubber or a blend of natural, SBR and BR have shoulder temperatures of around 120°C while running. Curing bags for tyre vulcanisation are made from butyl rubber and operate at temperatures around 160°C.

Silicone rubber has outstanding high-temperature properties and may even be used at temperatures up to 300°C for short periods. Inevitably the tensile strength of all rubbers falls as the temperature rises[7] as shown in Fig. 2. The best rubber with a carbon backbone is a fluorocarbon rubber but these are not very rubbery at room temperature.

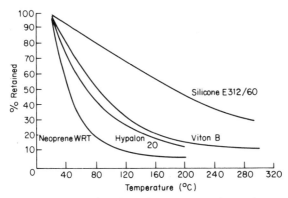

FIG. 2. Tensile strength as a function of temperature for various rubbers. (Reproduced from ref. 7.)

2.10. Ageing

It is important to distinguish between effects due to the crosslinking ingredients and those due to changes in the rubber itself. The simplest crosslinking system is that containing direct carbon–carbon crosslinks which are usually introduced by using organic peroxides. These peroxides react at a rate which is determined by the temperature so that if vulcanisation is carried out at a high enough temperature for a sufficiently long period of time all the peroxide is reacted and can take no further part. If some of the peroxide is left unreacted after vulcanisation then this will produce additional crosslinking during accelerated ageing tests at high temperatures.

Sulphur crosslinked rubbers show more complex behaviour. Conventional vulcanisates (high sulphur, low accelerator) usually show a maximum in degree of crosslinking as the cure time is extended. Prolonged exposure to high temperatures leads to a fall in the degree of crosslinking. This behaviour can be prevented to a large extent by the use of efficiently vulcanised sulphur systems (low sulphur, high accelerator) which show little change in degree of crosslinking once the maximum value has been reached.[1]

Of the unsaturated rubbers polychloroprene has the best resistance to atmospheric ageing but all the saturated rubbers such as ethylene propylene rubber have excellent ageing behaviour. Unfortunately the latter rubber cannot be crosslinked by sulphur since double bonds are absent, and peroxide curing systems present some problems. This difficulty is partly overcome by introducing a small number of crosslinkable units into the rubber chains but even so the rate of vulcanisation is relatively slow. Both silicone rubbers and fluorocarbon rubbers have excellent ageing behaviour although the latter is not very rubbery at room temperature.

2.11. Cost
The relative costs of various rubbers are shown in Table 1. Since costs are subject to considerable fluctuations the figures should be regarded as approximate. Also it should be remembered that the cost of the rubber is only one factor in the cost of the finished product.

2.12. Conclusion
It might be supposed that the properties of rubbers described above will exclusively determine which type is chosen for a particular application. However, this is not always so and the rest of this chapter is concerned with other factors which may enable the designer to have a much wider choice of rubber compound.

3. SHAPE EFFECTS

3.1. Oxidative Ageing
The shape of a component may influence its effective lifetime very dramatically if the component is in contact with an aggressive environment. Consequently tests on thin test pieces with a large surface area may give misleading information about the performance of a particular rubber compound if the actual component is much more bulky than the test piece.

A good example of this behaviour is found in conventional oven-ageing tests which are intended either to give information about the effect of high temperatures on the physical properties of the compound or as an accelerated test for the behaviour over much longer time periods at a lower temperature. Even if we confine our attention to the former aspect it is quite clear that the results are misleading.[8] Thin test pieces oven-aged at 120°C show dramatic changes in hardness, tensile strength, etc., after 7 or 14 days ageing whereas sections cut from the shoulder region of large truck tyres, which are known to have running temperatures of about 120°C, do not show such changes even after many months of use. The reason for this behaviour is that the ageing process which occurs in the laboratory specimens is due to oxidation from atmospheric oxygen, but the effect is confined to the surface layers only. In thin test pieces the depth of the oxidised layer may be a significant fraction of the thickness but in thick components it may be quite insignificant. The reason for this behaviour is that the oxygen present in the atmosphere diffuses into the rubber through the surface which is in contact with the atmosphere but as it diffuses through the rubber the oxygen reacts and thereby becomes used up. If the reaction rate is high compared with the diffusion rate then all the oxidative effects are localised near to the surface. The ageing which occurs in the bulk is anaerobic and quite different in character, being strongly influenced by the crosslinking system but not a serious problem with the appropriate crosslinking system.

The example just described demonstrates that ageing tests on thin test pieces do not simulate the effect on thick components even when they are operating at the same temperature as the test. It is not surprising therefore that when such tests are used to predict the behaviour at much lower temperatures (i.e. ambient) over a longer time scale the results are again misleading.

Probably the longest recorded in-service application is that of a natural-rubber sewer gasket which has been in use for more than 100 years.[9] The exposed surface was degraded to a depth of 2–3 mm but below this the rubber was still quite satisfactory. Another example of longevity is the Pelham Bridge bearings which are blocks of natural rubber that are still giving satisfactory performance after 25 years under loaded conditions and open to the atmosphere.

There is no evidence therefore that atmospheric oxygen causes properly compounded rubber components to deteriorate significantly. If components are thin and exposed to high temperatures then rapid failure may occur. Oxidation effects are catalysed by UV radiation and metallic ions so that it is possible for oxidation to become a serious problem if either of these catalysts are present. UV radiation is effectively absorbed by using small quantities ($<1\%$) of carbon-black filler so that there is only a problem with non-black rubbers. A much more serious problem for many rubbers (those containing double bonds) is caused by the presence of atmospheric ozone and this will be discussed later.

3.2. Oil Resistance

Another problem which is serious for thin components but which may be insignificant with thick components is oil resistance. Hydrocarbon oils are absorbed to a greater or lesser extent by all rubbers but those which are oil-resistant absorb a relatively small amount. A non-oil-resistant rubber may absorb up to twice its own volume of oil at equilibrium but the time taken to reach equilibrium depends on the viscosity of the oil and on the distance of the centre of the rubber from the surface in contact with the oil. The precise relationship will be discussed later. The total volume of the swollen rubber is equal to the sum of the volume of rubber plus the volume of the oil absorbed.

The physical properties of the swollen rubber such as tensile strength, tear strength and abrasion resistance are substantially reduced by the presence of the oil. The modulus and hardness are also reduced significantly. The oil does not dissolve the rubber, it simply mixes with it, but the strength properties are so reduced if the amount absorbed is large that the rubber may be rendered useless.

The equilibrium amount of oil absorbed is determined by the nature of the oil and the rubber as well as the degree of crosslinking and filler loading. Conventional swelling tests measure how much oil is absorbed by thin samples in a fixed time. However, the time recommended is often insufficient to ensure that equilibrium has been reached. Inspection of engine mounts and gear-box mounts on motor cars reveals that the rubber is often covered in oil but still performing satisfactorily. These mounts are usually made from a rubber which is not oil-resistant so it is clear that conventional swelling tests are not relevent to this application. The rate of penetration of the oil depends on its diffusion coefficient in the rubber rather than on the amount absorbed at equilibrium. The time taken for the oil to penetrate a given distance in natural rubber is easily calculated with the aid of the nomogram[10] shown in Fig. 3. The diffusion coefficient of the oil in the rubber may not be known but it has been found empirically that the viscosity of the oil is related to the diffusion coefficient as indicated on the nomogram. A low-viscosity oil penetrates more rapidly than a higher-viscosity oil. The same is true if the rubber is oil-resistant but, since the total amount of oil absorbed is small, the effect of oil viscosity is not usually important for these rubbers.

The following examples should make the use of the nomogram clear. A ruler is placed as shown by the dashed lines to show that an oil with viscosity of 300 cP would penetrate a distance of 1 mm in 4 weeks but would take 100 years to penetrate 4 cm. Although the nomogram refers to natural rubber it may be used for other rubbers with a similar glass temperature.

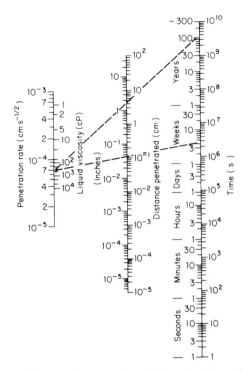

FIG. 3. Nomogram for rate of penetration of oils and solvents in natural rubber. (Reproduced from ref. 10.)

It should be emphasised that there is no rubber which is resistant to all liquids so it is necessary to find out which liquids may come into contact with the rubber. If other desired physical properties preclude the use of a rubber which is resistant to the liquid in question, efforts should be made to design the component so that the exposed area is small and so that the depth of rubber below the surface is as large as possible. A good example of limited access by a swelling liquid is given by rail pads which are flat pads sandwiched between the concrete sleeper and the shoe which holds the rail. The only exposed surfaces are the edges and hence the penetration of oil from these surfaces will be unlikely to affect the performance of the pad. In contrast, hoses have large areas of rubber in contact with the liquid which they are carrying and also they are thin so it is essential that the rubber is correctly chosen so that it is not swollen significantly by the liquid.

3.3. Heat Build-up

Heat build-up occurs as a result of rapid cyclic deformations of rubber. When a rubber is deformed, energy is stored by the rubber, but when it is allowed to recover, work is done by the rubber. However, the work recovered is always less than the work required to deform the rubber and this difference in energy appears as heat in the rubber. The heat is generated throughout the bulk of the rubber but cooling can only occur from the surfaces so that components with a considerable depth of rubber from the surface to the centre may be prone to generating high temperatures in the thickest part if the cyclic deformation is at a sufficiently high frequency. The problem is made more acute since rubber is a poor conductor of heat even when containing carbon-black filler at the levels normally used in engineering components (i.e. 40–60 parts by weight of black per hundred of rubber) or other high-quality products.

The ratio of the work recovered to the work required to deform the rubber is called the resilience R and is normally expressed as a percentage thus:

$$R = \frac{\text{energy returned}}{\text{energy supplied}} \times 100$$

The hysteresis H is simply $H = 100 - R$ if R is expressed as a percentage. The resilience of a compound is normally measured using a standard test, usually a rebound test of some sort. The resilience is a time-dependent property (i.e. depends on the rate of deformation) so that results on different machines are not usually comparable directly. Generally speaking the higher the speed of impact the lower the resilience if measurements are made at room temperature. However, the resilience passes through a minimum as the temperature is lowered and then rises as the temperature approaches the glass transition temperature[11] as shown in Fig. 4. The resilience at room temperature is therefore related to the glass transition temperature.

Rubbers with a low glass transition temperature, T_g, will have high resilience at room temperature whereas rubbers with a high T_g will have a low resilience at room temperature. There is an exception, butyl rubber, which has a low glass transition temperature but a very wide minimum resilience and the effect of this is still present at room temperature so that the resilience is still relatively low. The effect of fillers is to reduce the resilience at room temperature whereas increasing the degree

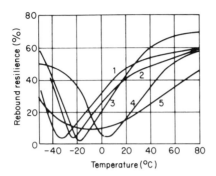

FIG. 4. Resilience as a function of temperature. 1, Natural rubber; 2, styrene/butadiene rubber; 3, chloroprene rubber; 4, butadiene rubber; 5, butyl rubber. (Reproduced from ref. 11.)

of crosslinking increases resilience and since both these factors increase hardness and modulus there is a reasonable amount of choice in the various combinations that may be made. High resilience compounds can only be made from well-crosslinked, unfilled rubbers with a low T_g. The resilience of rubbers can be increased by using a plasticiser but inevitably the strength of these compounds is reduced compared with a similar compound without the plasticiser. Often there is a loss in hardness but this can be offset by increasing the degree of crosslinking.

Heat build-up depends on hysteresis and resilience, but not in a simple way, since these properties only determine the heat-generation behaviour. The rate of cooling depends on thermal conductivity and the maximum thickness of rubber so that only general guidelines can be given. Heat build-up can be reduced by using a more resilient compound and by altering the shape so as to reduce the thickest part of the rubber. However, it should be remembered that if the purpose of the rubber is to act as a damper and therefore to dissipate energy then that energy will appear as heat and the methods suggested above will not only reduce the heat but also reduce its effectiveness as a damper. In this case a redesign of the shape is most important to ensure that adequate cooling occurs.

Although there is no cyclic deformation of the whole component in applications such as rollers and tyres, they are still subject to heat build-up problems because each element on the circumference is deformed as it enters the contact area and recovers as it leaves. Each element on the circumference therefore passes through an asymmetric deformation cycle (recovery period is long compared with the deformation period). With small tyres the maximum thickness of rubber is little more than a centimetre and there is no heat build-up problem but larger tyres (truck tyres and aircraft tyres) have to be made from a rubber with a low glass transition temperature to prevent the temperature rising to such an extent that blow-out occurs. The rubber used is either natural rubber or a blend of natural rubber and polybutadiene.

3.4. Surface Stresses

The life of components is often determined by the growth of cracks from flaws (invisible to the eye) in the surface.[12] Internal flaws are equally troublesome, if present, but usually only arise due to bad mixing or compounding. Surface flaws may arise from a variety of causes: accidental scratches and cuts, a damaged mould, ozone attack, or abrasion. None of these flaws matter if they are in a region where the surface stresses are compressive but they are a potential source of failure if the surface stresses are tensile since these stresses cause the cracks to open and consequently to grow. The rate at which cracks grow depends on a number of factors but the two most important ones are the type of rubber and the magnitude of the tensile stress around the crack.

Care should be taken when designing a component to avoid stress concentrations which occur at the edges and corners particularly if the rubber is bonded to a rigid plate of metal although the same principle applies to solid rubber as well.[12] Cracks grow preferentially in regions of high stress concentration and the avoidance of sharp corners increases the fatigue life enormously[12] as shown in Fig. 5.

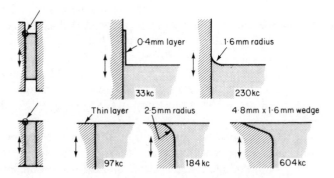

FIG. 5. Effect of reducing stress concentrations on the fatigue life of bonded natural-rubber components. (Reproduced from ref. 12.)

Ozone, which is present in the atmosphere at a concentration of only a few parts per hundred million of air, causes rapid cracking of unprotected rubbers containing carbon–carbon double bonds if the rubber is subjected to a sufficiently large tensile stress. The critical stress for cracking to occur depends on the rubber and it is quite low in unprotected rubbers. The inclusion of antiozonants either singly or in combination will raise the critical stress considerably in static applications but not in dynamic applications although in the latter case the rate of cracking is reduced considerably.[12]

A novel way of protecting components against ozone attack is to shape the surface in such a way that there are no tensile stresses.[13] This can be done for components designed to work in compression by making the exposed surfaces concave in the undeformed state. When the sample is compressed the rubber bulges outwards until the concavity becomes plane and, if compression continues, ultimately convex. While the surface is concave or plane it is under a surface compression and ozone does not cause cracking. A block which has been exposed to a very high concentration of ozone (50,000 × atmospheric) is shown in Fig. 6. In order to demonstrate the efficacy of the method, the block, which is made from unprotected natural rubber has two concave surfaces and two flat surfaces in the undeformed state. After compression and exposure to ozone for 24 h the concave surfaces are unmarked whereas the flat surfaces are badly cracked. The sample was compressed by 10% which caused the concave surfaces to become almost flat and the flat surfaces to become convex.

3.5. Shape Factor

The load-deformation behaviour of a rubber spring depends on the modulus of the rubber and on the shape of the spring. The modulus of the rubber is often measured in tension at 100% strain or even 300% strain but many technologists are more familiar with hardness. This is a measure of the modulus at low strains (i.e. Young's modulus) as shown in Fig. 7[14] and it is obtained using a standard instrument which measures the distance a hemispherical indentor is pushed into the rubber by a standard load.

The engineer is more often concerned with compressive and shear deformations at strains generally well below 25% in compression and 100% in shear so that to a good approximation the stress–strain behaviour is linear. The designer can therefore use the familiar shear modulus and Young's modulus in many cases, but a complication arises in

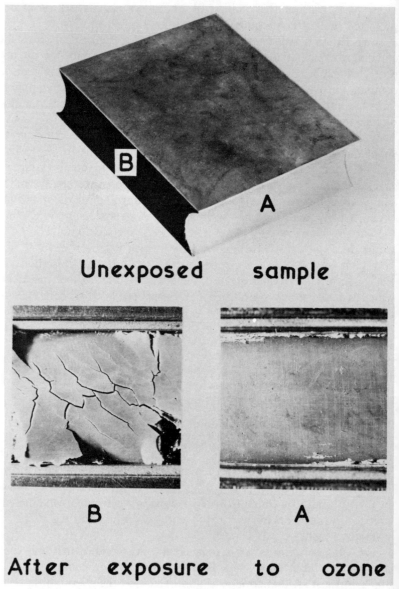

FIG. 6. Block exposed to ozone at high concentration showing effect of surface shaping to reduce ozone attack. (Reproduced by courtesy of MRPRA.)

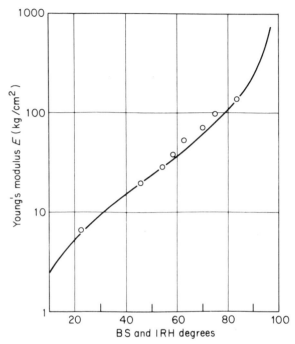

FIG. 7. Relation between Young's modulus and hardness. (Reproduced from ref. 14.)

the compression of flat slabs of rubber.[15] The compression modulus E_c depends on the shape of the block and is calculated from Young's modulus E_0 using

$$E_c = E_0(1 + 2KS^2)$$

where K is a slowly varying constant shown in Table 3 and S is the shape factor defined as follows:

$$S = \frac{\text{loaded cross-sectional area}}{\text{force-free area}}$$

For thinner pads the force-free area is smaller and therefore S is larger so that E_c is much larger than E_0. The reason for this behaviour is that rubber has a high bulk modulus and it can for most purposes be regarded as incompressible. For small strains Poisson's ratio is 1/2 but

TABLE 3
HARDNESS AND ELASTIC MODULI[16]

Hardness (IRHD)	Young's modulus E_0(MN m^{-2})	Shear modulus G(MN m^{-2})	Bulk modulus E_∞(MN m^{-2})	K
30	0·92	0·30	1000	0·93
35	1·18	0·37	1000	0·89
40	1·50	0·45	1000	0·85
45	1·80	0·54	1000	0·80
50	2·20	0·64	1030	0·73
55	3·25	0·81	1090	0·64
60	4·45	1·06	1150	0·57
65	5·85	1·37	1210	0·54
70	7·35	1·73	1270	0·53
75	9·40	2·22	1330	0·52

this falls to a much lower value at higher strains because of the incompressibility of rubber.[17] The shear modulus is not influenced by shape in this way so the ratio of shear modulus to compression modulus may be varied considerably by the shape of the block. This effect is widely utilised in applications where a low shear stiffness is required coupled with a high compression stiffness by the use of laminated blocks consisting of alternate layers of rubber and steel. The shape factor of the composite block of rubber can be increased by introducing more steel plates which reduces the thickness of each rubber layer and hence its compression modulus without affecting the shear modulus. Examples of this type of construction are seen in bridge bearings, building mounts, chevron bearings, and helicopter rotor bearings.

3.6. Protective Coatings

Proprietary paints are available to coat rubber components either to give them an aesthetically attractive finish or, more often, to protect the rubber from a hostile environment. The paints are usually rubber-based (e.g. polychloroprene, chlorosulphonated polyethylene, urethane) since they have to be capable of sustaining the same sort of deformations as the rubber component. Usually they are needed to give protection against oil, or ozone, or acids, so the appropriate rubber-based paint is chosen. While the paint film will reduce the rate of swelling of the underlying non-oil-resistant rubber it will not affect the ultimate degree of swelling. The film would have to be completely impermeable to the oil in order to prevent the oil from reaching the rubber below. Nevertheless the rate can be substantially reduced and indeed it is a relatively simple matter to measure the rate at which the oil is transmitted through the membrane and to calculate therefore the time taken for the underlying rubber to absorb any particular quantity of oil. It is essential that the adhesion between the paint and the rubber is not destroyed by the oil. A major disadvantage is that if the paint film becomes ruptured for any reason protection around the rupture is lost.

Protection against ozone attack will be mainly in reducing the rate unless the film is completely impermeable to ozone or unless it contains an antiozonant which will react with the ozone which is permeating the film. Again, a rupture of the film will mean a complete loss of protection in the damaged area.

A coating which is continuously being formed and therefore is self-healing if scratched is produced by the blooming of waxes to the surface of rubbers. This is a widely used method of protecting against ozone attack in static applications. The wax film breaks up if the rubber is deformed very much. The wax is mixed into the rubber during processing at a level which exceeds the solubility at ambient temperatures (the solubility depends on the particular rubber and wax used but it is usually around $0.5-1.5\%$). During vulcanisation, which occurs at temperatures greater than the melting point of the wax, the latter is completely soluble but on cooling to room temperature a supersaturated solution of rubber and wax is formed which usually results in the wax being precipitated in the form of particles in the bulk of the rubber. The wax in these particles is gradually transferred to the surface of the rubber, thereby producing the protective film.

The mechanism of blooming is a little complicated involving the gradual redissolution of the wax particles as a result of the elastic pressures from the rubber surrounding them and then diffusing to the surface of the rubber.[18] Eventually all the wax will bloom to the surface but this normally takes several years. The rate of blooming is not linear with time being much more rapid in the early stages than at later times. The waxes are crystalline materials, hence their low solubility below their melting point and therefore they should confer some protection against oil and solvents which will only be slightly soluble in them. However, this possible application does not appear to be supported by experimental observations which are recorded in the literature.

3.7. Surface Treatments

The surface properties of rubbers can be changed by reaction with a suitable chemical reagent. Chlorine and bromine, usually dissolved in water, are effective in chlorinating or brominating the rubber. Either of these treatments makes the surface of the rubber more polar and hence improves the effectiveness of some adhesives. The halogenated surface has lower friction than the untreated surface while the bulk is unaffected, so this method can be used to reduce the friction coefficient where an easy sliding action is desirable (e.g. seals and windscreen-wiper blades). In addition, the hard shiny surface is aesthetically attractive and this, together with the lack of surface tack, makes it ideal for such items as rubber handgrips or handles.

Another surface treatment is cyclisation. This occurs when rubber is dipped in concentrated sulphuric acid. This is a more severe treatment than halogenation and it produces a hard brittle surface which gives improved performance with some adhesives.

3.8. Surface Finish

Apart from aesthetic considerations, the surface finish is important in so far as it should not be the source of surface flaws. All failure processes result from the growth of flaws or cracks in the rubber. The larger the flaw, the faster it grows so the rate of growth accelerates as the crack grows. It is important in these circumstances that the size of any flaws in the surface should be as small as possible. It appears that there are natural flaws within the rubber which are about 1/100 mm in size,[19] so that extra care in preparation does not improve this figure. If larger flaws are introduced by inadequate dispersion of fillers, large dirt particles, or by surface flaws or scratches, these will grow much faster than the natural flaws and cause failure to occur more readily.

3.9. Type of Deformation

There are three basic types of uniaxial deformation: tension, compression and shear. More-complicated deformations can be resolved into these three basic forms. From the foregoing discussions it is clear that tensile stresses should be avoided if at all possible since this type of deformation leads to the growth of cracks. Compressive stresses, in contrast, do not cause flaws to open and therefore no growth occurs. It should be remembered, however, that in general a compressive stress may lead to a surface tensile stress unless the surface is contoured as described earlier. Shear stresses have both compressive and tensile elements along the principal axis so that cracks will develop perpendicular to the major principal axis.

Components are generally designed to work in compression and shear in order to minimise the possibility of cracking. Often shear deformations are combined with compression so that the components are 'fail safe' even if they crack or become unbonded as shown in Fig. 8.

Some satisfactory components have been made where the deformation is mainly tensile, notably in motor-car exhaust-pipe suspensions. These are essentially a circular hoop of rubber with no bonding to metal. By making a suitable choice of rubber compound, ozone attack is minimised so that an acceptable lifetime, in excess of 5 years, is obtained. Another application is in furniture webbing or seat diaphragms where again a lifetime of several years is easily achieved by suitable compounding. However, it must be admitted that, although these designs are satisfactory, it is unusual to find rubber components deformed in simple tension. Finally, it is worth noting that the stress–strain properties of rubber are linear over a range of strain values up to 25% in compression, depending on shape factor, and up to 80% in shear, but not in tension.

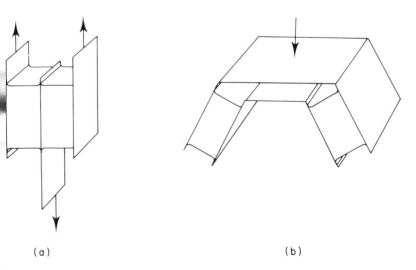

FIG. 8. (a) Double shear component and (b) fail safe combined compression and shear.

An important factor which must be remembered in any design is that rubber is not compressible, its bulk modulus is very high compared with the shear modulus[17] (a factor of between 100 and 1000). It is this feature which leads to the shape-factor effects already discussed. The smaller the area of free surface compared to the volume of the rubber the stiffer it will be in compressive deformation. This is particularly important in sealing applications using seals such as O-rings. The groove must be large enough to accommodate the bulging of the ring as it is compressed. The manufacturers literature for these seals gives the dimensions of the groove which is required to allow for this effect.

3.10. Fatigue Life

The fatigue life of a rubber component is determined by the growth of small flaws as a result of cyclic stresses. These flaws are initially invisible (about 1/100 mm) but as they grow they eventually become large enough to be seen easily. However, the rate of growth increases with increasing crack length so that by the time the cracks are easily visible the life is almost over.[19] Cracks also grow in some rubbers as a result of static loading. This behaviour is in addition to the effects of ozone which causes cracking in unsaturated rubbers (those with double bonds) as described earlier.

It is convenient to divide rubbers into two categories, those which strain-crystallise and those which do not. Strain-crystallisation occurs when rubbers which have a regular molecular structure are highly strained. These rubbers also crystallise even when unstrained if kept at low temperatures for a long enough time. The flaws in rubbers of this type only grow while they are being stretched. If the stretching is insufficient to cause rupture during the straining process then the cracks cease to gow once stretching ceases.

Non-strain-crystallising rubbers behave quite differently; the cracks continue to grow in these rubbers even when the straining action has ceased but the rubber is still strained. In a cyclic deformation the strain-crystallising rubbers also benefit greatly if the strain cycle does not pass through zero strain.[12] The fatigue life can be extended by a hundredfold simply by ensuring that the minimum strain is above 50%. No such benefits occur with non-strain-crystallising rubbers. With these rubbers increasing the minimum strain merely shortens the life since cracks grow all the time that the rubber is strained. The importance of this effect should not be underestimated since its magnitude is much greater than that which can be achieved from compounding changes. The latter can give improvements of, at best, less than a factor of ten compared with the hundredfold improvement which can be achieved by raising the minimum strain in the case of strain-crystallising rubbers.

The rate at which the cracks grow depends on the particular rubber and on the maximum stress applied. An example of the behaviour is shown in Fig 9 where the abscissa is not stress as might be expected but tearing energy.[12] This parameter has been found to be more fundamental a quantity in strength properties than the applied stress although it is related to the stress in addition to the test-piece dimensions and, depending on the type of test, on the stress–strain properties of the rubbers.[20] The tearing energy is the work which must be done to cause a crack or flaw to grow and to create unit area of new surface. It is much greater than the surface energy of the rubber (by a thousandfold). This is due to the work done in deforming the rubber in the region of the growing crack. The strains close to the tip of the growing crack are

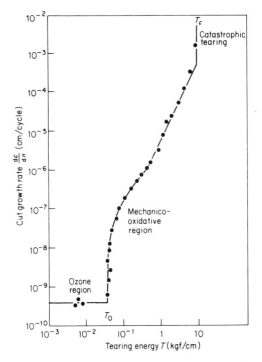

FIG. 9. Crack-growth characteristic of natural rubber. (Reproduced from ref. 12.)

very large, almost up to the breaking strain, and these strains are reduced to a very low value after the crack are very large, almost up to the breaking strain, and these stains are reduced to a very low value after the crack has advanced. The hysteresis in the rubber plays an important part in dissipating energy, thereby making it more difficult for the crack to propagate.

Rubbers with a lot of hysteresis or compounds with reinforcing fillers which increase hysteresis have superior performance in reducing the rate at which cracks grow in the presence of a stress.

4. CONCLUSION

It should be clear from the foregoing that there are many factors to be taken into account when designing a product. Some of these arise from the specification, others are at the discretion of the designer. In all cases it is important to be aware of the advantages and disadvantages associated with a particular rubber compound so that a correct assessment can be made of the viability of the compound for the particular application. If this is not done one aspect may be improved but the component may still fail prematurely because another equally important aspect has been overlooked. When mistakes of this sort occur rubber products generally are condemned whereas it is the component designer who is to blame. Well-designed rubber products are capable of a longevity which compares favourably with most engineering materials and in addition they are generally maintenance-free. It can not be emphasised too strongly that conclusions drawn from one application should not be used to forecast what will happen in a quite different application. An extreme example of this would be a stationer's rubber band which performs quite satisfactorily the job for which it is designed (i.e. holding envelopes together for a short period of time) but would fail catastrophically in a few weeks if left stretched outdoors. The same rubber (i.e NR) compounded differently and deformed in compression is still perfectly satisfactory after 21 years in service supporting the Pelham Bridge

REFERENCES

1. SOUTHERN, E. In *Elastomers: Criteria for Engineering Design*, C. Hepburn and R. J. W. Reynolds (Eds.), Applied Science Publishers, London, 1979, Chapter 16.

2. TAYLOR, G. R. and DARIN, S. R. *J. Polymer Sci.*, **17**, 1955, 511.
3. MULLINS, L. In *Chemistry and Physics of Rubber-like Substances*, L. Bateman (Ed.), Applied Science Publishers, London, 1963, Chapter 11.
4. ANDREWS, E. H. and GENT, A. N. In *Physics and Chemistry of Rubber-like Substances*, L. Bateman (Ed.), Applied Science Publishers, London, 1963, Chapter 9.
5. FLETCHER, W. P., GENT, A. N. and WOOD, R. I. *Proc. 3rd Rubber Technol. Conf., London*, 1954, 382.
6. SMITH, J. F. *Rubber in Engineering Conf., Kuala Lumpur*, MRPRA Reprint 764, 1974.
7. THOMAS, D. K. and SINNOTT, R. *J. IRI*, **3**, 1969, 163.
8. RUSSELL, R. M. *Brit. Polymer J.*, **1**, 1969, 53.
9. SMEE, A. R. *Rubber Developments*, **16**, 1963, 115.
10. SOUTHERN, E. In *Use of Rubber in Engineering*, P. W. Allen, P. B. Lindley and A. R. Payne (Eds.), Maclaren, London, 1967, Chapter 4.
11. MULLINS, L. *Trans. Inst. Rubber. Ind.*, **22**, 1947, 235.
12. LAKE, G. J. and LINDLEY, P. B. In *Use of Rubber in Engineering*, P. W. Allen, P. B. Lindley and A. R. Payne (Eds.), Maclaren, London, 1967, Chapter 5.
13. DERHAM, C. J., SOUTHERN, E. and THOMAS, A. G. *Intern. Rubber Conf., Moscow*, 1969; *NR Technol.*, **1** (7) 1970.
14. GENT, A. N. *Trans. Inst. Rubber Ind.*, **34**, 1958, 46.
15. GENT, A. N. and LINDLEY, P. B. *Proc. Inst. Mech. Engrs*, **173**, 1959, 111.
16. LINDLEY, P. B. *Engineering Design with Natural Rubber*. NR Technical Bulletin, MRPRA, Hertford, 1974.
17. LINDLEY, P. B. *Bull. Mech. Eng. Educ.*, **6**, 1967, 186.
18. SOUTHERN, E. *Diffusion of Liquids in Rubber*, Ph. D. Thesis, London University, 1970.
19. GREENSMITH, H. W., MULLINS, L. and THOMAS, A. G. In *Physics and Chemistry of Rubber-like Substances*, L. Bateman (Ed.), Applied Science Publishers, London, 1963, Chapter 10.
20. RIVLIN, R. S. and THOMAS, A. G. *J. Polymer Sci.*, **10**, 1953, 291.

INDEX

Accelerators, 105, 116, 122–5, 142
  developments, 117–18
  fast-curing, 124
Activated sulphenamide, 125
Additives, 69, 188–9
  dispersion-promoting, 187–8
Adhesion, 198
  brass, to, 202–4
  EPDM, to, 204
  tests, 202
Ageing, 259–60
Alkoxysilanes, 189
Aminofunctional silane, 196
Aminophenol groups, 14
Antidegradants, 228
Antioxidants, 228
  chemical reaction with rubber, 242
  consumption statistics, 229
  efficiency, 237–9
  high-molecular-weight, 242
  non-staining, 236–9
  permanence
    effects, 241–5
    improvement, 242–5
  selection of, 245
  staining, 232
  synergism, 240–1
Antiozonants, 228, 267
  commercially available structures, 234
  consumption statistics, 229
  non-staining, 239–40
  staining, 232–6

Aromatics, 210
ASTM Iodine Number, 158
Azo 'ene' reaction, 37

BET equation, 157
Bismaleimide system, delayed-action, 24
Bisphenol cure systems, 78–81
Bladder cancer, 246
Blending, 240
Blooming, 271
Brass, adhesion to, 202–4
Bridge structure, 238
Bromination, 272
Bulk modulus, 274
Bundesgesundheitsamt, 246
Butyl rubber, petroleum oils in, 223
Butyl titanate, 189
Butylated hydroxytoluene (BHT), 232

Carbon blacks, 151–82, 184
  application areas, 176
  basic features, 151–62
  bonding compounds, 170
  Bueche's mechanism, 161
  bulk-handling, 179
  car-tread compounds, 169
  casing compounds, 171
  channel, thermal, acetylene and gas-furnace processes, 164
  characterisation, 156

Carbon blacks—*contd.*
  compound performance, 168–75
  correlation index, 166
  effectiveness factor, 160
  electron micrograph, 153
  energy loss/temperature generation, 173
  formation, 151–3
  future prospects, 176–9
  grade distribution, 177
  hysteresis, 161
  improved, 162–3
  inner liner, 171
  lampblack process, 163–4
  low rolling resistance tyres, 172
  nature of, 154–6
  new products, 178–9
  new technology, 162–3
  non-tyre applications, 172–3
  oil-furnace process, 165, 176
  pellet properties, 179
  production processes, 163–6, 186
  properties, 154–6
  quality
    assurance, 167
    control, 166
  rationalisation, 175
  reclaimed, 179
  reinforcement mechanism, 160–2
  run-flat tyres, 171
  sidewall compounds, 171
  structure, 156, 158, 168–9
  surface area, 155, 168–9
  truck-tread compounds, 170
  tyre applications, 169–72
  ultimate properties, 161
Carborane–siloxane polymers, 97–8
Carboxy nitroso rubbers, 96–7
Cetyl trimethylammonium bromide (CTAB), 158
Chlorinated polyethylene, 51–3
  applications, 53
  compounding, 52
  grades, 51
  preparation, 51
  properties, 52
  vulcanisation, 52
Chlorination, 272

Chlorosulphonated polyethylene, 47–51
  applications, 50–1
  basic compounding, 47–50
  basic properties, 47
  chemical structure, 47
  crosslinking mechanisms, 50
  grades, 47
  manufacture, 47
  vulcanisation, 47–50
Coal tars, 208
Composites, interrelationships in, 186
Compounding effects, 215–20, 227
Compounding oils, 209
Compressibility, 274
Compression, 272, 273
  modulus, 269, 270
Cost aspects, 260
Coupling agents, 185, 194
Coupling mechanism, 186
Cracks, 266, 272, 275–6
Creep, 256
Crosslink density in SBR/BR, 147
Crosslink distributions, 126, 145
Crosslink reactions v. curing temperature, 144
Crosslink stability, 148
Crosslink structure, 112–15
Crosslinking, 252, 256, 259, 262, 265
  agents, 19, 254
Crystallisation, low-temperature, 254
Curing system for equal modulus, 136
Cyclisation, 272
N-Cyclohexylbenzothiazole-2-sulphenamide (CBS), 11

Deformation types, 272–4
De Mattia test, 220
Deproteinised grade of natural rubber (DPNR), 7
Design. *See* Product design
Dialkyl-*p*-phenylenediamines, 233
Diamine cure systems, 77–8
Diaryl-*p*-phenylenediamines, 236
2,2′-Dibenzothiazyl disulphide (MBTS), 11, 22
Dicumyl peroxide, 21, 23, 24

INDEX 281

Dilauryl thiodiproprionate (DLTP), 241
N,N'-Di-β-naphthyl-p-phenylenediamine (DNPPD), 245, 246
N,N'-Diphenylguanidine (DPG), 11

Ethylene–vinyl acetate (EVA)—contd.
  chemistry, 89
  compounding, 90
  grades of, 90
  properties, 90
Extenders, 208, 253

Efficiency parameter, 113
Elastic modulus, 173, 270
'ene' reaction, 37
ENPCAF, 37, 38
Epichlorhydrin, 53
  elastomers (CO ECO), 53–7
    applications, 57
    basic properties, 54
    chemistry, 53–7
    compounding, 55–6
    grades, 55
    preparative methods, 54
    raw polymer properties, 54
    structures, 54
    thermal degradation, 56
    vulcanisation, 55–6
Epoxidation of natural rubber, 40–2
Esters, 209
6-Ethoxy-2,2,4-trimethyl-1,2-dihydroquinoline, 233
Ethylene acrylic elastomer, 61–5
  applications, 64–5
  chemistry, 61
  compounding, 62
  properties, 62, 65
  vulcanisation, 62–4
Ethylene oxide, 53
Ethylene/propylene copolymer (EPM), 222, 241
Ethylene/propylene/diene modified rubber, 33, 222, 239, 240
  adhesion to, 204
  peroxide-cured, 194, 200
  sulphur-cured, 196, 200
  vulcanisation, 141–2
Ethylene–propylene polymers, 222
Ethylene–vinyl acetate (EVA), 89–91
  applications, 91
  availability, 89

Fatigue life, 266, 274
  natural rubber, 18, 25, 137
  SBR, 137–9
Filler/rubber bonds, 185
Fillers, 64, 69, 82, 253, 262, 264, 276
  catalytic effect, 191–3
  classification, 187
  cure inhibition, 191
  effect of additives, 188
  effect of water, 187
  energy requirements, 184
  fracture, 190
  matching, 194–6
  pretreatment, 189–90
  reinforcement, 198
  rheology, 186–90
  rubber compounds, in, 194
  surface modification, 185
  viscosity, 188–9
  see also Carbon blacks; Silane-treated mineral fillers
Flexibility, low-temperature, 255
Fluorocarbons (CFM and FKM), 74–85
  applications, 85
  chemistry, 74–5
  commercial sources, 75–7
  compounding, 81–2
  fluids resistance, 83–4
  grades, 75–7
  low-temperature properties, 84–5
  properties, 82–5
  vulcanisation, 77–81
Fluorosilicones (FVMQ), 72–4
  applications, 74
  chemistry, 72–3
  grades, 73
  properties, 73–4
Food and Drug Administration, 246

INDEX

Glass transition temperature, 255, 264
Grafting effects, 243

Halogenation, 272
Hardness, 251, 253, 254, 256, 269, 270
Health aspects, 245–6
Heat
  build-up, 264–6
  resistance, 258
Hevea brasiliensis, 6
Hexamethylene diamine carbamate, 77
High-temperature curing, 143–7
High-temperature properties, 258
Hindering group effects, 238
Hysteresis, 264, 276

Injection moulding, 143

Kelvin model, 173

Lampblack process, 163–4, 173
Liquid elastomers, petroleum oils in, 225
Load-deformation behaviour, 267
Low-temperature behaviour, 254, 255

Malaysian Rubber Producers' Research Association (MRPRA), 242, 243
Marching modulus, 138
2-Mercaptobenzimidazole (MBI), 241
2-Mercaptobenzothiazole (MBT), 11
Mono-sulphidic crosslinks, 9
Moore–Trego efficiency parameter, 113

Naphthenics, 210
β-Naphthylamine, 246
Natural rubber, 1–44, 240
  ageing and reversion resistance, 134

Natural rubber—*contd.*
  applications, 3, 24–9
  chemical modification, 35–7
  crosslink distributions, 126, 145
  crystallisation, 2–3
  cure testing, 6–7
  delayed-action peroxide vulcanisation, 19–23
  deproteinised, 7
  effect of carbon black, 26–9
  effect of curing system on overcure 131
  effect of curing temperature on tensile strength, 135
  'ene' reaction, 37–40
  engineering properties, 24–9
  epoxidised, 40–2
  EV systems, 24, 130–6
  fatigue life, 18, 25–6, 137
  general-purpose, 2, 6
  highly epoxidised, 42
  loss factor, 26–9
  modern forms of, 3–7
  modification, 33–42
  molecular structure, 3
  oil-extended, 29–32
  petroleum oils in, 222–3
  physical properties, 15
  premasticated, 32
  processing and curing characteristics, 135
  quality of, 3
  retreading compounds, 31–2
  semi-EV system, 130–6
  soluble-EV system, 11–13, 24
  strength and tack of, 3
  stress–relaxation rate, 26–9
  thermoplastic blends, 33–5
  types and grades, 4
  tyres, 29–32
  versatility, 3
  viscosity-stabilised grades, 4–6
  vulcanisation systems, 8–23, 130, 136
Newtonian flow, 186
Nitrile rubber (NBR), 244
  petroleum oils in, 225
  vulcanisation, 140–1

Nitroso rubbers, 96–7
N-Nitrosodiphenylamine, 22
p-Nitrosodiphenylamine (NDPA), 242
Norbornene elastomer
  applications, 99
  availability, 99
  chemistry, 98–9
  compounding, 99
  properties, 99

OENR compounds, 29–31
OESBR compounds, 29–31
Oil
  penetration nomogram, 262–3
  resistance, 257–8, 261–3
Organic acids, 209
Organic peroxides, 19
O-rings, 274
Oxidation effects, 260–1
Oxidative heat ageing, 11
N-Oxydiethylenebenzothiazole-2-
    sulphenamide (OBS), 11
Ozone attack, 227, 267, 271

Paints, 270
Paraffinics, 210
para-substituent effects, 238
Peel strength, 199, 200
Perfluorinated elastomer (PFE), 86–9
  applications, 89
  availability, 86
  chemistry, 86
  properties, 86–9
Peroxide
  cure systems, 80
  vulcanisation, delayed-action, 19–23
Petroleum oils, 209–14
  analysis, 211
  butyl rubber, in, 223
  carbon-type analysis, 212–14
  composition effect, 219–20
  compounding effects, 215–20
  ethylene–propylene polymers, in, 222
  liquid elastomers, in, 225
  loading effect, 217

Petroleum oils—contd.
  molecular-type analysis, 211–12
  natural rubber, in, 222–3
  nitriles, in, 225
  polychloroprenes, in, 223
  processability, 215–17
  stability, 221–2
  types of, 209–11
  viscosity, 218
  vulcanisate properties, 217
Phenols
  bridged hindered, 237
  simple hindered, 236–7
Phenyl-α-naphthylamine (PAN), 228, 245, 246
Phenyl-β-naphthylamine (PBN), 228, 232, 245, 246
N,N'-m-Phenylenebismaleimide, 22
p-Phenylenediamine(s), 228
  derivatives, 232, 233–6
N-Phenyl-N'-substituted p-
    phenylenediamine, 242
Phosphonitrilic fluoroelastomers, 95–6
Pine-tar products, 208
Plasticisers, 64, 82, 207–25, 255
  definition, 208
  types of, 208–9
Plasticity retention index, 4
Poisson's ratio, 269
Polyacrylic elastomers (ACM), 57–61
  applications, 61
  chemistry, 57–8
  commercial sources, 58
  compounding, 60
  grades, 58
  processing, 60
  properties, 60
  vulcanisation, 58–60
Polyaromatic hydrocarbons, 159
Polychloroprenes, petroleum oils in, 223, 236, 240
Polydichlorophosphazene, 95
Polydimethylsiloxane (MQ), 66
Polyethylene, 51
Polyisoprene, 2–4
Polymethylmethacrylate, 35
Polyolefin, 34
Polypropylene, 33, 34

Polystyrene, 40
Polysulphide rubbers (T), 91–5
  applications, 95
  chemistry, 91
  compounding, 94
  hydroxyl-terminated, 91
  mercaptan-terminated, 92–3
  properties, 93–4
Polysulphidic crosslinks, 9
Prevulcanisation inhibitors, 121
Primers, 203–4
Process aid, 208
Product design, 249–77
  general requirements, 249–51
  selection of rubber compound, 251–60
  shape effects, 260–76
Protective agents, 227–47
Protective coatings, 270–1

Radical scavenger, 20–2
Refractivity intercept, 213
Reinforcement concepts, 185–6, 198
Resilience, 264, 265
Retarders, 120–1
Retreading compounds, 31–2
Rheometer
  test, 6–7
  traces, 20
Rubber consumption, statistics, 228–32

Seals, 274
Shape factor, 267–70
Shear
  modulus, 267, 270, 273, 274
  stresses, 272
Silane-treated mineral fillers, 183–205
  adhesion, 196–8
  effects on rubber cure, 192–3
  performance, 194–6
  recommendations for usage, 198
Silicone elastomers, 65–72
  applications, 72
  chemistry, 65–7
  commercial sources, 67

Silicone elastomers—*contd.*
  compounding, 68–9
  grades, 67
  heat-curable compounds, 68, 70
  heat-vulcanisable polymers, 66
  high strength, 70
  MQ designation, 66
  properties, 69–72
  room temperature vulcanising (RTV) compounds, 67–9, 72
  VMQ designation, 66
  vulcanisation, 67–8
Silicone fluid, viscosity changes, 189
SMR-CV, 6, 24
SMR-GP, 6
SMR-LV, 6
Solvents
  penetration nomogram, 262–3
  resistance to, 258
Special-purpose elastomers, 45–103
Stabilisation, 232
Stabilisers, 232, 246
Standard Malaysian Rubber (SMR) scheme, 4
Stearic acid, 12, 209
Strain-crystallisation, 274–5
Stress
  concentrations, 266
  relaxation, 12, 26–9, 256
Stress–strain properties, 267, 273
Styrene/butadiene rubber (SBR), 244
  crosslink distributions, 126
  EV systems, 136
  fatigue life in, 137–9
  processing and curing characteristics, 140
  properties of, 139
  semi-EV systems, 136
  silica-filled formulations, 193
  sulphur-cured, 200
  vulcanisation, 136–40
Styrene/butadiene/styrene (SBS), 241
Sulphenamide, 143
Sulphur donors, 118–19
Sulphur-soap cure systems, 60
Sulphur/sulphenamide system, 23
Sulphur vulcanisation. *See* Vulcanisation

surface
  finish, 272
  stresses, 266–7
  treatments, 272
swelling, 270–1
Synthetic elastomers, 45
Synthetic rubber, 244
  production, 232

Tackifiers, 203–4
Tensile strength, 251–3, 256, 258
Tensile stresses, 267, 272, 273
Tetrabutylthiuram disulphide (TBTD), 11
Tetraethylthiuram disulphide (TETD), 11
Tetramethylthiuram disulphide (TMTD), 11
Tetramethylthiuram monosulphide (TMTM), 15
Thiazole, 125, 143
Toxicity aspects, 245–6
Tributyl thiourea, 239
Trimethylolpropanetrimethacrylate, 21
Tris(nonylphenyl)phosphite (TNPP), 241
Tyres, natural rubber, 29–32

Urethane(s), 147
  reagents, 13–19
Urethane/sulphur systems, 15, 18, 19

Vibrational damping, 65
Viscoelastic behaviour, 256
Viscosity/gravity constant, 212
Vulcanisate structure and properties, 10

Vulcanisation, 8–23, 105–49, 254
  ACM, 58–60
  CM, 52
  CO, 55–6
  continuous, 143
  conventional systems, 122
  CSM, 47–50
  ECO, 55–6
  effects on vulcanisate properties, 111
  EPDM, 141–2
  ethylene acrylic rubber, 62–4
  EV system, 8–11, 114, 125–42
  EVA, 90
  fluorocarbons, 77–81
  NBR, 140–1
  non-sulphur, 147–8
  NR, 130, 136
  parameters of, 110
  peroxide, delayed-action, 19–23
  SBR, 136–40
  selection of system, 115–21
  semi-EV systems, 129–42
  silicone elastomers, 67–8
  soluble-EV systems, 11
  sulphur, 8–11, 107–11
  sulphur/sulphenamide, 109
  urethane reagents, 13–19
  see also High-temperature curing

Waxes, 240, 271

Young's modulus, 9, 267

Zinc dimethyldithiocarbamate (ZDMC), 15
Zinc 2-ethyl hexanoate, 12
Zinc soaps, 12